EVOLUTION OF
LIVING ORGANISMS

040436
26·9·78

£13.85

3/94

UNIVERSITY

EVOLUTION OF LIVING ORGANISMS

Evidence for a New Theory of Transformation

PIERRE-P. GRASSÉ

Université de Paris VI
Laboratoire d'Évolution des Étres Organisés
Paris, France

ACADEMIC PRESS   New York   San Francisco   London   1977

A Subsidiary of Harcourt Brace Jovanovich, Publishers

First published in the French language under the title "L'Évolution du Vivant" and copyrighted in 1973 by Éditions Albin Michel, Paris, France.

ACADEMIC PRESS, INC.
111 Fifth Avenue, New York, New York 10003

United Kingdom Edition published by
ACADEMIC PRESS, INC. (LONDON) LTD.
24/28 Oval Road, London NW1 7DX

Library of Congress Cataloging in Publication Data

Grassé, Pierre Paul
    Evolution of living organisms.

    Translation of L'évolution du vivant.
    Bibliography: p.
    Includes index.
    1. Evolution.   I.   Title.
QH366.2.G713   575   77-80786
ISBN 0-12-295550-1

Contents

III
Evolution—A Discontinuous Historical Phenomenon

IV
Evolution and Chance

V
Evolution and Natural Selection

VI
Evolution and Adaptation

VII
Evolution and Necessity

VIII
Activities of the Genes in Relation to Evolution

IX
A New Introduction of Evolutionary Phenomena

Appendixes

Preface

Some biological problems are equally appealing to the scientist and philosopher. The evolution of living organisms is one of them. We should rejoice at such a meeting of interest and curiosity, but the philosopher does not have an accurate, extensive, and "assimilated" knowledge of existing data. This may also be said of the biochemist, geneticist, or laboratory worker who lacks direct contact with nature, the essence of evolution in action. Nevertheless, the philosopher's participation in evolutionary studies is not to be underestimated. He boldly plunges into the field of metaphysics where the naturalist "fears to tread."

I would like to think that this book serves both the needs of the biologist and the philosopher. In reality, though, to write such a book exceeds the scope of a biologist, which is all I claim to be. Yet, I believe the philosopher will find in this work a source of thought and criticism of not only my ideas but those of my illustrious predecessors.

In this book I have made an effort to present only those propositions that have been well established by scientific study or directly observed in my own experiments. Half a century of research in various disciplines of zoology and general biology has given me some insight into the realities of the living world.

A proper understanding of evolutionary phenomena requires a thorough knowledge of zoology, paleontology, cytology, genetics, biochemistry, and even mathematics. No one could hope to master such a wide range of scholarship. Yet without it how can evolution be treated pertinently? "Book knowledge" is not enough. A practical, concrete knowledge of fossils and of living animals, preferably observed in their natural environment, is of prime importance to the evolutionist. Who, having seen a seal or walrus swimming, would dare to claim that these animals survived merely because chance led them to adapt to an aquatic habitat? One need only observe a giant Brazilian otter swimming with unrivaled grace and ease to realize that (of course simplifying somewhat) an evolutionary process that started with a carnivorous fissiped of amphibious habits gradually led to pinnipedal carnivores restricted to living in water.

Any account of evolution that loses sight of paleontologic data is but a theory dominated by the imagination. It would seem that paleontologists, zoologists, and botanists have not yet recovered all the information stored in the paleontologic record. It is likely that a fresh survey, conducted with a fully open mind, would bring to light unknown facts and orient research into yet unexplored realms.

Many of the ideas expressed in this book will seem disconcerting to the English or American reader schooled in orthodox Darwinism. With this caveat, let him overcome initial reluctance and read the book. He will, I maintain, discover the unavowed weakness of a doctrine that falls far short of universal explanation. To guarantee a well-founded knowledge of evolutionary phenomena and to demonstrate the failure of currently held doctrine to account for them is no vain enterprise. I consider it to be a supremely useful and fundamentally scientific one.

In this book, I attempt to prove that the traditional interpretations assigned to evolution are not the only ones offered the biologist or natural philosopher. I propose one which, while also based on paleontologic and molecular biological data, aspires to give a rational and general account of real evolutionary phenomena.

My sincere thanks are due Professor Bruce M. Carlson and Ms. Roberta Castro, without whose efforts this book could not have been published in English. They accepted the task of revising my English translation to eliminate language problems, a difficult task in which they have succeeded admirably.

Pierre-P. Grassé

EVOLUTION OF
LIVING ORGANISMS

An Introduction to the Study of Evolution

GENERAL OBSERVATIONS ON THE LIVING WORLD

To say that living beings are composed of the same elements as inert bodies, that they are subject to the same chemical and physical phenomena, is in fact extremely commonplace and rather meaningless.

How could the composition of living beings be different when there exists a single multiform matter? What would their properties be if they did not depend on physicochemical laws? Would they be ruled by Melusina's magic wand or by Merlin's incantations? Such suppositions are absurd.

To know that living beings share with material bodies the same basic chemical composition and obey the same physicochemical laws does not teach us anything fundamental about the originality of life.

One of the radical differences between physical and biological phenomena lies in the fact that the former must necessarily and absolutely obey the laws of matter. Gravity exerts its action on all material bodies; nothing in the gravitational field escapes its universal influence. The living being reacts to physical law, escaping from it to a greater or lesser extent. For instance, the wings of birds and insects allow them to defy the law of gravity without violating it.

Necessity does not imperatively impose its laws on the living world. Proof of this is seen in the morphological and functional variety of plants and animals that succeed in overcoming the greatest physical difficulties, living in polar or torrid zones; they manifest, within the same environment, a great diversity in form and behavior. Therefore, the living being, because of its structural complexity, its mechanisms, and

1

its "inventions"* partly escapes physical laws or eludes them. One of its constant victories is indeed to avoid the law of entropy and to become a "machine" which permanently opposes it.

Every living being obeys its own law, which is to remain as it is and to produce new beings identical to itself. Oak trees remain oak trees (in themselves and in their offspring). Only another law, the specific law of the line, makes the living being violate its own law, forces change upon it, and pushes it into the cycle of biological evolution.

The living world is thus governed by rules to which the Universe of inert matter is not subject. We rarely discover these rules because they are highly complex; they are not easily expressed in mathematical terms because the number of parameters they involve is so great.

The calcium carbonate, silicates, plaster, iron, copper, silver, and gold which form St. Peter's Basilica in Rome do not differ essentially from the same materials found in the mines and quarries of the four continents, and yet how different they are. The layout, plan, protective devices, decoration, and adaptation to a given purpose have all come into play; all this was conceived by man, created by his mind, without which there would be nothing but "raw" material.

Any living being possesses an enormous amount of "intelligence," very much more than is necessary to build the most magnificent of cathedrals. Today, this "intelligence" is called "information," but it is still the same thing. It is not programmed as in a computer, but rather it is condensed on a molecular scale in the chromosomal DNA or in that of any other organelle in each cell. This "intelligence" is the *sine qua non* of life. If absent, no living being is imaginable. Where does it come from? This is a problem which concerns both biologists and philosophers, and, at present, science seems incapable of solving it.

When we consider a human work, we believe we know where the "intelligence" which fashioned it comes from; but when a living being is concerned, no one knows or ever knew, neither Darwin nor Epicurus, neither Leibniz nor Aristotle, neither Einstein nor Parmenides. An act of faith is necessary to make us adopt one hypothesis rather than another. Science, which does not accept any credo, or in any case should not, acknowledges its ignorance, its inability to solve this problem which, we are certain, exists and has reality. If to determine the origin of information in a computer is not a false problem, why should the search for the information contained in cellular nuclei be one?

---

*The use of this term in biology has been forbidden by some, but there is no better word to express adaptations. Since I avoid talking about transcendant powers, I use *"invention"* without metaphysical connotation, expressing only an actual fact.

The powers of invention in the living world are immense. In our opinion, they are nothing but the capacity to process information in a given direction and perhaps toward a given goal. We do not know their inner mechanism and their underlying sources: biologists grope in darkness.

Plants and animals themselves build the organs which bring them into contact with the outside world and enable them to survive or improve their existence. Such construction, achieved by continuing efforts over tens of millions of years, does not surprise us: it seems so common, so natural to us, and yet what a huge difference from nonliving matter!

Has a stone ever acquired anything other than a crust of decomposition? The physicochemical laws which it obeys prevent its acting independently.

It is, of course, true that living beings are subject to the laws common to all matter, but they also obey certain rules which, although still dependent on physicochemical laws of a particular type, are specific to them alone. Both types of law complement each other, and if discord occurs between them, the living being dies. Life results from the adjunction of complementary systems to some determined material systems, the former being likewise determined but specific to organized beings. Niels Bohr's principle of complementarity applies not only to the corpuscular and vibratory aspects of light, but also to living beings. One of the essential tasks of biochemists and biophysicists is to recognize in living beings what belongs to the nonliving and what belongs to the living.

BIOLOGICAL EVOLUTION AND ITS OCCURRENCES

The term "biological evolution" refers to the succession and variation in time of plant and animal forms. It implies that to parental continuity be added an internal tendency to modify certain structures and to create new ones. Zoologists and botanists are nearly unanimous in considering evolution as a fact and not a hypothesis. I agree with this position and base it primarily on documents provided by paleontology, i.e., the history of the living world.

The study of contemporary species does not establish the existence of evolution; it provides facts which support it, but which do not fully demonstrate its existence. This is understandable, since at present we cannot show the series of successive stages which make up evolution, but only a fleeting picture of evolution.

To maintain that structures, functions, and behaviors can only be understood if they take part in an evolutionary process is a method to which I sometimes have recourse. It is not altogether reliable since it postulates implicitly that our logic is that of the living world, of which we have no certainty.

Naturalists must remember that the process of evolution is revealed only through fossil forms. A knowledge of paleontology is, therefore, a prerequisite; only paleontology can provide them with the evidence of evolution and reveal its course or mechanisms. Neither the examination of present beings, nor imagination, nor theories can serve as a substitute for paleontological documents. If they ignore them, biologists, the philosophers of nature, indulge in numerous commentaries and can only come up with hypotheses. This is why we constantly have recourse to paleontology, the only true science of evolution. From it we learn how to interpret present occurrences cautiously; it reveals that certain hypotheses considered certainties by their authors are in fact questionable or even illegitimate.

Embryogenesis provides valuable data to anyone who knows how to interpret it with circumspection and subtlety. Ernst Haeckel's fundamental biogenetic law, despite overwhelming criticisms by various biologists, has not lost any of its value. Its author has been credited with opinions that he never held. He perfectly understood that ontogeny is indirectly influenced by environmental adaptations and changes, although it exhibits a great resistance to variations.

It is not certain that a planktonic larva, such as the trochophore of annelids, of mollusks and of other allied groups, represents an ancestral state; it is possible that it is only a state adapted to the dissemination of the species and inscribed in its genetic code. However, it can be inferred from its structural plan that annelids and mollusks have fairly close affinities and probably common, although remote, ancestors.

The panchronic species, those relics which preserve ancient forms and are the vestiges of extinct faunas, throw light not only on evolution as such, but also on its mechanism. Biology has not yet recovered all the relevant information that they contain in the history of the animal kingdom and on the mechanisms of evolution.

Certain chemists, wanting to find in both kingdoms facts proving evolution and revealing the relationships between the various groups of plants or animals, have undertaken some highly interesting studies. For example, phosphagens which occur in great quantity in muscles (up to 0.5% of the muscle mass) are the main "reservoir" of phosphoric links, releasing great energy available for the work of muscular contraction. In invertebrates, the phosphagen is phosphoarginine, whereas in vertebrates and prochordates it is phosphocreatine. However, annelids possess a particular phosphagen (phosphoglycyamine or phosphotaurocyamine), and some echinoderms possess phosphocreatine (*Ophiura*) or a compound of phosphoarginine or phosphocreatine (Urchins). Moreover, differences in the enzymes catalyzing the chain of reactions are added to these differences in the chemical composition.

Certain serological and immunological reactions and the determination of "specific proteins" allow us to define with increasing accuracy the limits of systematic groups. Chemistry, through its analytical data, directs biologists and provides guidance in their search for affinities between groups of animals or plants. It makes presumptions supporting given interpretations, and in this respect plays an important part in the approach to genuine evolution. It is to be deplored, however, that biochemists, finding the same substance or observing the same reaction in distinct groups, too often decide on the existence or relationships which are contrary to zoological and paleontological data.

The discovery of fundamental biochemical processes common to all living beings (the citric acid cycle, for example) has reinforced the monistic conception of the two kingdoms. Until 1970, molecular biology had not really contributed anything new to evolution; information on genes, genetic mutations, and on chromosomes was new. The discovery of the molecular structure of nucleic acids and of the genetic code has allowed a better understanding of the nature and action of the gene. But genetics is the science of "heredity," of the preservation of the specific gene pool; its relationships to evolution are known only through theories.

INTERPRETATION OF EVOLUTIONARY FACTS

It is surprising to note that present-day theories explaining evolution are based on the same principles as in the past. This practice has in no way weakened the claims of the advocates of these theories; on the contrary, many Anglo-Saxon and a few French biologists so strongly believe in these doctrines that they write without the slightest hesitation that the evolutionary mechanism is known in detail with a high degree of certainty. We have gone from Darwinism into neo-Darwinism, and, very recently, to ultra-Darwinism, which not only claims to be the sole custodian of truth in regard to evolution, but to be evolution itself. Darwin himself did not display so much confidence when writing to one of his grandsons: "But I believe in natural selection, not because I can prove in any single case that it has changed one species into another, but because it groups and explains well, it seems to me, a lot of facts in classification, embryology, morphology, rudimentary organs, geological succession and distribution...."*

Present-day ultra-Darwinism, which is so sure of itself, impresses incompletely informed biologists, misleads them, and inspires fallacious interpretations. The following is one of the numerous examples found in books today: "In microorganisms, the generation time is rather short and the size of the population can be enormous. Therefore, *mutation acts as a very powerful evolutionary process* during a shorter lapse of time than in populations of higher organisms" (Lévine, 1969, p. 196, the italics are mine). This text suggests that modern bacteria are evolving very quickly, thanks to their innumerable mutations. *Now, this is not true.* For millions, or even billions of years, bacteria have not transgressed the structural frame within which they have always fluctuated and still do. It is a fact that microbiologists can see in their cultures species of bacteria oscillating around an intermediate form, but this does not mean that two phenomena, which are quite distinct, should be confused; the variation of the genetic code because of a DNA copy error, and evolution. *To vary and to evolve are two different things;* this can never be sufficiently emphasized, and we will try to prove that this proposition is correct later in the book.

Bacteria, which are both the first and the most simple living beings to have appeared, are excellent subject material for genetic and biochemical study, but they are of little evolutionary value.

Through use and abuse of hidden postulates, of bold, often ill-founded extrapolations, a pseudoscience has been created. It is taking root in the very heart of biology and is leading astray many biochemists and biologists, who sincerely believe that the accuracy of fundamental concepts has been demonstrated, which is not the case. Wishing to point out this type of misconception, we quote P. T. Mora, an American biochemist, who writes about polysaccharides contained in the cell membrane: "Of course we know that such specific structure is the result of the working of enzymes, which in turn is a reflection of the genetic information transmitted by nucleic acids through cycles of reproduction as *selected by evolution* (Mora, 1965, p. 40, the italics are mine). To admit that the action of enzymes and, more important, that their formation is directed by the genetic code should not permit one to maintain that the information was selected by evolution (the consequence is mistaken for the cause); no one knows anything about this. Today, the evolutionary

---

*This letter is in the British Museum. It was published with a facsimile by Vernet (1950).

value of selection is measured with little more accuracy than at the time Darwin wrote the above-mentioned letter.

For those who would still doubt the relevance of our criticisms, another quote from a paper written by two American biologists, King and Jukes (1969), is presented:

> Darwinism is so well established that it is difficult to think of evolution except in terms of selection for desirable characteristics and advantageous genes. New technical developments and new knowledge, such as the sequential analysis of proteins and the deciphering of the genetic code, have made a much closer examination of evolutionary processes possible, and therefore necessary.

These biologists are convinced that the Darwinian credo is correct, and they accept it. They are quite sincere but they are not critical enough.

Biochemists and biologists who adhere blindly to the Darwinism theory search for results that will be in agreement with their theories and consequently orient their research in a given direction, whether it be in the field of ecology, ethology, sociology, demography (dynamics of populations), genetics (so-called evolutionary genetics), or paleontology. This intrusion of theories has unfortunate results: it deprives observations and experiments of their objectivity, makes them biased, and, moreover, creates false problems.

Heedful of genetics and demography, Darwinians have seldom taken fossils* into consideration, or, and this is more serious, they have applied the laws of genetics to them without making a critical analysis; considering our ignorance of the relationships between fossils, which in most cases are found very far apart and in distinct beds, this approach can only be arbitrary. Paleontologists, who cannot have recourse to experiments when deciding that a given character is genetically valuable, thus expresses a very hypothetical opinion. Assuming that the Darwinian hypothesis is correct, they interpret fossil data according to it; it is only logical that they should confirm it: the premises imply the conclusions. The *error in method* is obvious.

Assuredly, Darwinism holds a number of trumps, the highest ones being its simplicity and its logic. It also poses as an exact science by applying statistics to evolution, through the use of genetics, without first

---

*This is so true that in his classic work, a compendium of neo-Darwinism, Huxley (1942) lists only 24 titles of publications having reference to a greater or lesser degree to fossils, whereas the bibliography includes 890 titles. Theoretical interpretation of variations undergone by modern animals and plants predominates over consideration of fossil forms; yet these are the only forms in which evolution has materialized.

asking the fundamental questions: Are the postulates used scientifically exact? Is the use of mathematics really necessary?

Lamarckism, which is no less logical than Darwinism, but in an entirely different perspective altogether, considers life a system self-reacting to the influence of the environment and capable of harmonizing its functions with them. This is a tempting theory. But as the heritability of acquired characters has not yet been proved experimentally, it is hard to adopt it as a doctrine explaining evolution. Wintrebert (1962) has tried, with some success, to give it a chemical basis by involving the capacity of every living being to react to environmental changes and aggressions. What Wintrebert lacks is an experiment corroborating this. All has not been said, however, and we would not be surprised to learn from molecular biology that some of its intuitions are partly true. Although Lamarckism draws strongly favorable arguments from the structures and functions of present and fossil beings, it should be considered today a way of thinking, of understanding nature, rather than a strict doctrine entirely oriented toward the explaining of evolution.

The code of conduct that the naturalist wishing to understand the problem of evolution must adopt is to adhere to facts and sweep away all *a priori* ideas and dogmas. Facts must come first and theories must follow. The only verdict that matters is the one pronounced by the court as proved facts. Indeed, the best studies on evolution have been carried out by biologists who are not blinded by doctrines and who observe facts coldly without considering whether they agree or disagree with their theories. Today, our duty is to destroy the myth of evolution, considered as a simple, understood, and explained phenomenon which keeps rapidly unfolding before us. Biologists must be encouraged to think about the weaknesses of the interpretations and extrapolations that theoreticians put forward or lay down as established truths. The deceit is sometimes unconscious, but not always, since some people, owing to their sectarianism, purposely overlook reality and refuse to acknowledge the inadequacies and the falsity of their beliefs.

It is true that, with regard to evolution, it is not easy to have access to reality; the past does not lend itself easily to our research, and experiments do not have any hold over it. What is completed leaves traces, but escapes from our intervention. The evolutionist is therefore always in search of the past. His quest is difficult, but not hopeless.

I

From the Simple to the Complex—Progressive Evolution, Regressive Evolution

Evolution implies the filiation of matter and many other conditions. It does not result from random, incoherent variation, occurrring without order. In reality, it proceeds in a continuous and orderly manner, in line with trends which become more and more pronounced with the passing of generations; it thus establishes lines of descent and large families. Evolutionary trends are so numerous that they leave the path open to diversity.

If the variations of plants and animals were not subject to rules, their appearance and structure would change somewhat as a star glitters or as corpuscles move at random under Brownian motion. *Evolution would not exist.* The impression of disorder which sometimes follows a superficial study of past and present faunas fades away when forms are studied in depth, when chronology is explicitly defined, and when evolutionary lines are established. If the variation of the living world were nothing but chaos and disorder, it could not possibly be the subject matter of a science. At the very best, statistics could be applied to it.

Paleontology reveals the existence of lines, defines their tendencies, analyzes their features, and, by measuring the various degrees of evolution, assigns to each genus, to each species, its corresponding rank. Paleontology even goes one step further: it reveals certain characteristics common to several lines, and thereby establishes affinities.

Paleontology became a science the day Cuvier, noticing the jaw of a mammal imbedded in a block of gypsum, studied its features and predicted the structure of the other bones and even announced the existence of marsupial bones.* Only the predictable can belong to an order. Paleontology exists solely because chaos is banned from the living world. One sees the keenest scorners of oriented evolution proceed in Cuvier's manner and spend the greatest part of their time discovering natural lines, bringing to light systematic affinities between animals, and drawing up family trees. However, when they conduct their research as paleontologists, they forget their refutations and theories and work in a constructive manner.

If one considers animals and plants, from the most remote times up to the Tertiary or Quaternary era, it appears that the structures of the species have undergone a great number of processes and have become considerably more diverse. What was the extent of this elaboration and what was its effect on the transformation of the species?

Although the term "progressive" evolution is not inappropriate, we will seldom use it, because it expresses a more or less hidden and often meaningless moral judgment. Progress, in biology, implies the existence of two successive stages, the second stage prevailing over the first; for the second stage to prevail, it must produce some feature which is advantageous to the individual, and to the species. According to Darwinians, species "progress" because the individuals which have undergone favorable variation prove themselves more capable of leading a certain form of life, of performing a specific function, and because they have the most numerous offspring. The "unvarying" individuals will be eliminated more or less rapidly, depending upon the intensity of the "selection pressure" which bears upon the population to which they belong. This pressure is sometimes so intense that all the invariants die.

Natural selection *directs* evolution for the best interest of the species: the individual that has evolved, the fittest according to Darwinian theory, is progressive compared to the "unchanged." This proves that the word progressive expresses a moral judgment and assigns a goal to evolution to proceed toward the best.

INCREASING COMPLEXITY AS A FUNCTION OF TIME

As soon as living beings appeared in the seas of our planet they exhibited a clear dual trend: toward complexity and diversity, the second being, in a way, complementary to the first. A simple comparison between the uniformity of the flora of the first millions of years (consist-

*Paired bones articulating with the pubis and anteriorly free, present in the wall of the pouch of monotremes and marsupials.

ing solely of bacteria and blue-green algae), and the superabundance of the thallophytes of the Tertiary era is proof of the diversified rising of the species. It is a common opinion that biological evolution never ceased proceeding from the simple to the complex. Hence the hypothesis that the simplest living forms were the first to appear in the primitive ocean, and that their fossilized remains are found in the most ancient sediments of the earth's crust.

Biologists all agree that bacteria and Cyanophyceae, or blue-green algae, are the living beings with the simplest structure. They are lumped together as the division Schizophyta, characterized by the lack of a readily identifiable, condensed nucleus (no nuclear membrane) and the lack of chromosomes (their DNA strand together with proteins does not make up a specific organelle), of mitochondria, and of ergastoplasm. They are further characterized by reproduction by binary fission and the possibility of partial sexuality without total mixing of genetic pools.

In a few privileged areas of the globe, the scarce Precambrian sediments that were not destroyed by metamorphism contain fossils of a microscopic size; these have recently been identified with a great degree of certainty.

In the oldest formations, cherts of the Fig Tree formation of Barberton Mountain land (Eastern Transvaal), Barghoon and Schopf (1966) found traces of organic matter and a few microfossils. Measurements of the age of these rocks, by the radioactivity (loss) of strontium and rubidium, indicate that they might have been formed by sedimentation approximately 3.2 billion years ago!

Organic matter occurs in irregular "trails" of an undefined structure, up to 9 μm long. These trails could be granules composed of amino acids, used as nutrients by the bacteria, their contemporaries. The organic beings found in those cherts could be small bacteria, such as *Eobacterium isolatum*, a minute rod (less than 0.7 μm long, and with a 0.2 μm diameter) enclosed in a membrane (0.15 μm thick). It has been suggested that they utilized amino acids or even the proteins present in the seas. One can logically imagine that chemotrophic bacteria (iron bacteria) appeared later; they were autotrophic because their food ("prebionts") had disappeared due to the change of environmental conditions both in and out of water.

Some spheres, measuring 17 to 20 μm in diameter (*Archaeosphaera barbertonensis*), which were found in the Fig Tree formation, might be Cyanophyceae, i.e., schizophytes containing chlorophyll (the agent of photosynthesis which occurs with a release of oxygen through the decomposition of water); this is how a "breathable" atmosphere started to take form around the earth.

Other Precambrian cherts, less ancient than those of the Fig Tree formation (2.3 billion years old), found in the Gunflint formation (located in the vicinity of Lake Superior, in Western Ontario), have provided several plant fossils (8 genera and 12 species described); some of these (filamentous) fossils are related to some Cyanophyceae which resemble the present *Oscillatoria* and *Rivularia;* several structures either simply spheroidal, or possessing a central body from which branched and unbranched filaments radiate, have no equivalent in the present flora. Filamentous bacteria (up to several hundred microns long), apparently related to our iron bacteria (of the genus *Crenothrix*), have been found in the same sediments.

Kakabekia, an umbrellalike microorganism, which was first found in these sediments, was encountered alive in several soils (Alaska, Iceland, Hawaii); it is a schizophyte of uncertain phylogenetic affinities.

A few rather complex spheroidal organisms may be algae but they apparently do not have any descendants among present species. Moreover, chemical analysis has revealed that the sediments of the Gunflint contain pristane and phytane, two hydrocarbons which are thought to have been derived from the chemical alteration of chlorophyll; for this reason, they are often classified as chemical fossils.

In the Precambrian unmetamorphosed rocks (limestone, sandstone, dolomite) of the Amadeus basin (Central Australia: Bitter Spring formation), which is thought to be about one billion years old (absolute age), Schopf (in Barghoorn, 1971) found numerous microfossils comprising three species of bacteria, 20 blue-green algae, two unquestionable green algae, and, in addition, two fossils interpreted with some doubt as fungi. These various fossils, which are from 3.2 to 1 billion years old, give evidence of two essential stages in evolution:

1. The *prebiotic stage*—which is revealed by the organic, amorphous trails; these may have been the substances from which the first organic beings were formed, and at the same time their source of nutriment.

2. The *genesis of schizophytes* (bacteria and Cyanophyceae)—chlorophyll first appeared on the planet with Cyanophyceae.*

With regard to the cellular state, it is likely to have occurred in algae during the upper Precambrian era approximately 1 billion years ago.

What were the evolutionary steps from the *"schyzophyte"* structure (that of bacteria and Cyanophyceae) to the true cell structure (i.e., a nucleus enclosed in a membrane and containing chromosomes with a

*It is likely that the formation of pigments (agents of photosynthesis) occurred several times during the history of the biosphere, in blue-green algae, in chlorobacteria, and in green cellular plants.

well-defined structure, in which are enclosed DNA molecules combined, in some unknown way, with filamentous proteins)? We have no
idea. Paleontology does not reveal anything on this matter. Hypotheses
are plentiful, but they all lack substantiation.

There is no evidence of a continuity between unicellular algae and
multicellular algae. The same species sometimes goes through a regular
cycle involving both a unicellular and a multicellular stage. This is why
botanists confidently classify most protophytes among algae (where the
diversification of cells and the formation of organs remains limited). On
the animal side, the link between the uni- and multicellular organisms is
still missing. In spite of extensive study, the origin of the Metazoa is still
unknown.

For several reasons, and because of the differences in the modes of cell
division, the ancestry of Metazoa cannot be assigned to any group of
present Protozoa. A number of biologists now tend to consider Protozoa
as the descendants of protophytes lacking pigment, and Metazoa as
those of multicellular colorless algae (see Grassé, 1952, 1969). Given the
lack of fossil evidence, only tentative and extremely hypothetical solutions can be formulated for this problem. Paleontology provides evidence that, from the very beginning, the evolutionary trend was toward
complexity and expansion; what we mean is that it never stopped producing new and more complex bodies. It is essential to give due consideration to these fundamental characteristics of evolution.

If evolution has been supplied at random with the materials it needed,
it would have been unpredictable and the distribution of the species in
time would have been unspecific, disorderly, and chaotic. The fact is
that paleontologists observe just the contrary and can predict with assurance the order of the genesis following a mode of increasing complexity. Thus, the Precambrian fauna—mainly known from the fossils of the
Edicara sediments (Australia) and the fossils of a few beds in South
Africa, England, Colorado, and California—is as predicted, and it consists mainly of sponges, Cnidaria, and various other forms of uncertain
affinity, which are described later. At Ediacara, in the sands of shallow
waters, lived animals similar to the modern *Pennatula*, while jellyfish
and what are claimed to be Siphonophora floated above. These floating
species were so numerous that the period during which they were fossilized has been called the Era of the Jellyfishes.

There is a relationship between the degree of complexity of an animal
and the date of its appearance. As Saint-Seine (1951) wittily said: "Fossils were punctual at their meeting with forecasts." The fact that evolution followed a given calendar means that it obeyed certain laws which
biologists and paleontologists must define.

CHRONOLOGICAL ORDER OF APPEARANCE (Fig. 1)

One of the fundamental stages of biological evolution has been the passage from unicellular to multicellular forms. The simplest and most archaic multicellular beings are the sponges and Cnidaria. Their structure, whatever their external appearance, consists fundamentally of a central cavity surrounded by two layers of cells; hence the word *diploblastic* used to name them. Their cells, during ontogenesis, undergo differentiation into a number of different categories of varying specialization (collar cells, poisonous cells) but do not form any clearly individualized organs. In colonial species, the individual or zooid specializes in the performance of a specific function; it is either a float, a gastrozoid

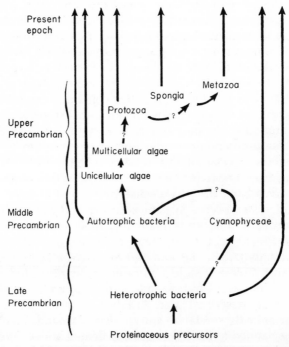

Fig. 1 Phyletic relationships among the various groups of living beings at the beginning of evolution. Other types of relationship have been conceived (see Grassé, 1962). The hypothesis that two kinds of cellular organisms would have arisen from bacteria is not absurd but seems unlikely. The first group would be bacteriophagous protozoans (of the *Bodo* type, for example) deriving from heterotrophic bacteria; the second group would be autotrophic Protophyta deriving from autotrophic bacteria. The uniformity of the structural plan and of cell organelles in both kingdoms makes a double origin of the cell rather unlikely.

(nutritive), a gonozooid (reproductive), or a dactylozooid (tactile and protective). The inability to acquire organs is not in contradiction with the increasing complexity of forms, as illustrated by hexactinellids among sponges, and siphonophores and actinoids among the Cnidaria; their cavity, always with a single opening, is the fundamental structure of diploblastic organisms.

Among sponges, nerves and sensory cells are incompletely differentiated. The specific feature of the nervous system is the synaptic connection of neurons with cells; however, readily identifiable synapses do not seem to exist in sponges. Lamarck rightly called the sponges, Cnidaria, and Ctenidia "apathetic"; their sensory equipment is indeed very basic. Moreover, if no specific area of their body has a musculature that enables them to perform rapid and coordinated movements.

The anatomy of the Cnidaria is more complex than that of the sponges; it consists of individualized nerve and muscular cells. No one knows, however, whether the Cnidaria and Ctenaria (both diploblastic organisms) are derived from the sponges, or whether they all stem from a common ancestor.

The chronological order of appearance of the phyla of triploblastic invertebrates is still unknown, as we have no fossil evidence of their primitive forms.

The acquisition of the third cell layer, mesoderm, gave metazoans extensive new potentialities, and in particular the ability to form individualized organs for the performance of specific functions. Among several triploblastic organisms, including the vertebrates, experiments have shown that induction does occur; it is produced either by the mesoderm or some of its variants or by the mesenchyme, and it plays a major part in organogenesis. There is no doubt that, as animals reached the triploblastic stage, they made a major step toward increased complexity. All triploblastic organisms, except for echinoderms which are so peculiar in many ways, exhibit a marked body "regionalization," including a "cephalization" that aggregates most sensory and nervous centers upon and in the anterior region of the body. The external information is now centralized in the brain where it is processed. Animals gained autonomy when they acquired a third cell layer, from which originated the muscles for rapid contraction and, in vertebrates, the internal bony skeleton.

The differentiation into specialized organs does not alter the unity of the being since connections, performed namely by the nervous system and the endocrine glands, make all the organs interdependent. Thus, the individualization of the living organisms becomes increasingly marked and their behavior greatly advances as they make use of ever-

increasing information of the external world (information that is processed more acutely by the central nervous system).

Once mesoderm appeared, the structural plans began increasing in
number and complexity. First appeared new phyla of invertebrates with
increasingly complex structural plans. Unfortunately, paleontology reveals little about their origin and even less about the first stages of their
history. It is most certainly from one of these phyla that the chordates,
which comprise the prochordates and the vertebrates, were derived.
Several zoologists consider the echinoderms to be the ancestors of chordates (metazoans with a skeletal rod, above which is located the tubular
nervous system, and under which extends the digestive system, of
which the anterior end or pharynx performs a respiratory function).
Their opinion is based on anatomical and embryological facts. The
paleontologist Jefferies, both alone (Jefferies, 1968) and in collaboration
with Prokop (Jefferies and Prokop, 1972), maintains that the chordates
stemmed from a group of very ancient echinoderms (the calcichordates)
that lived during the Cambrian and Ordovician periods. The body of the
calcichordates was enclosed in a theca, formed by united calcareous
plates; from a mineralogical standpoint, each plate is a single crystal,
which is a characteristic of the higher echinoderms. The mouth was
located at the front end of the theca; the internal cavities were the buccal
cavity, the pharynx, the anterior and posterior coeloms, and occasionally a few other chambers. The pores studding the theca were gill clefts.
The body extended posteriorly with a tail or pedicle, through which ran
a spinal cord. The brain lay in the anterior end of the peduncle, behind a
pyriform ganglion, and sometimes behind traces of cranial nerves.

Jefferies' (1968) detailed study is illustrated by very suggestive figures;
the reconstruction of the brain is shown very clearly. Since we have not
seen the original parts, it is impossible to distinguish the real from the
imaginary. In particular, it is doubtful whether the formation located in
the axis of the peduncle is really a chord as Jefferies contends. Jefferies'
entire interpretation is disregarded and considered as mere conjecture
by an authority on fossil echinoderms (Ubaghs, 1971), who writes: "As it
is obvious that nothing is or will ever be known with certainty about the
soft parts of extinct organisms, the basic hypothesis cannot be verified."
Our position is somewhat more encompassing than Ubaghs' (indeed,
the number of fossil soft parts which are known is not negligible). We
believe, however, that it would be unwise to accept any organ reconstruction and Jefferies' bold ideas without serious reservations.

The prochordates, probable ancestors of the vertebrates, left very few
fossils (as their bodies were soft and unprotected). The Permosoma

(Permian of Sicily) are related to the present Chelysoma (asidians), and spicules of Didemnidae (ascidians) shed no light upon the origin of the phylum.* The order of appearance of vertebrates is as follows: agnathans, fishes,† amphibians, reptiles, mammals.‡ This phylum is characterized by continuously increasing overall structural complexity as well as a steady increase in "psychism." This fact is supported by fossils.

As mentioned above, the prochordates (ascidians, *Amphioxus*) announced the vertebrates in some ways. Evidence for this is also plentiful. The structural plan of an ascidian tadpole, of an appendicularian, of an *Amphioxus,* although in a simpler form, is fish, and even of an amphibian, the patterns of the prospective mesodermal parts show resemblances which seem to indicate much more than simple analogy.

Accurately naming the group of invertebrates from which the phylum Chordata arises is another matter. Some zoologists hold that the metamerism of prospective mesodermal structures and the relationship of the excretory system and the gonads with the mesodermal cavities (coelom) relate vertebrates to annelids. These conclusions are no doubt very impressive but they do not bridge the immense gap between polychaete annelids and *Amphioxus.* How many missing links are there between the two of them? Nobody can tell.

We find it unnecessary to survey the other hypotheses put forward by phylogenists; they are not satisfactory since they are not based on any paleontological data.

Since we hardly know anything about the major types of organization, suggestions, and suggestions only, can be made. How can one confidently *assert* that one mechanism rather than another was at the origin of the creation of the plans of organization, if one relies entirely upon imagination to find a solution? Our ignorance is so great that we dare not even assign with any accuracy an ancestral stock to the phyla Protozoa, Arthropoda, Mollusca, and Vertebrata. The lack of concrete evidence relative to the "heyday" of evolution seriously impairs any transformist theory. In any case, a shadow is cast over the genesis of the fundamental structural plans and we are unable to eliminate it.

Ainiktozoon of the Upper Silurian, *Scaumenella* of the upper Devonian may possibly be prochordates. *Jamoytius* (upper Silurian) appears to be an anaspid agnathan and not a prochordate. These fossils are poorly preserved and can only yield questionable evidence (see Scourfield, 1937; White, 1946).

†Fishes form a superclass consisting of several major lines which are very distinct and have had various fortunes.

‡Mammals and birds were derived from reptiles, but their respective ancestors belong to orders which are distantly related.

Increasing organic complexity, the highest degree of which can be found in the anatomy and cytoarchitecture of the nervous system and sense organs, was followed by a concomitant blooming of the psychic faculties, leading to more and more precise, varied, and plastic behavior.

All motile living beings exhibit a behavior which is adapted to their needs and insures their survival. This property struck Ernst Haeckel (1877) and led him to assign to the cell a "psychogenic faculty" that, to his mind, was as essential as assimilation and reproduction. The father of materialistic monism, the apostle of atheism, even goes much further. In his answer to Virchow (Haeckel, 1882 French translation, p. 84) he writes: "Isolated single cells exhibit the same manifestations of psychic activity—feeling, perception, will, movement—that can be found in multicellular beings"; he even talks about the *"spirit" of the cell*. Haeckel's conclusions are based on unquestionable evidence. It is a fact that, during the course of evolution, "psychism" (a vague term, but which does include all types of behavior) continuously expanded and developed as a result of the elaboration and specialization of the nervous system.

In spite of their apparent simplicity, amebalike protozoans do react to light, heat, and chemical substances (acids and bases) and to mechanical perturbations; they also "choose" and reject unsuitable food. Ciliate infusoria show a more elaborate behavior still; they acquire, for short periods, conditioned reactions, and some spines of their membrane are said to perform sensory functions.

Neurility and muscularity thus appear in a rudimentary state in protozoans. In metazoans (except for sponges) these functions are performed by specialized cells, neurons and muscle fibers that are the result of a slow evolution.

Behavior depends upon the structure of the nervous system. For example, the structure of the nervous system of diploblastic organisms (all of which are acephalous and lack a brain) is so simple that these "apathetic" beings were classified for more than a century among the plants. True "cephalization" only starts to appear in triploblastic organisms and attains its highest grade of construction in arthropods and vertebrates, animals in which behavior is most complex.

In arthropods, the brain is divided into anatomically and functionally specialized parts; behavior, although refined, is nonetheless automatic. The earth structures and industrious activities of insects, whether they be social or not, result from closely related stereotyped acts. Instinctive complexes command and set off inborn mechanisms, the determining factors of which have been inscribed into their chromosomal DNA, as in the case of anatomical and functional characters.

With the multicentered brain of vertebrates, behavior is increasingly "freed" from genetic mechanisms and commands. Behavior acquires plasticity but still remains mostly inborn; it only becomes entirely liberated in man, whose evolution and brain have their own characteristic features (Grassé, 1971).

From the above evidence it follows that evolution, considered globally, has followed a path of increasing structural complexity, accompanied by a constant increase in "psychism": automatic (culminating in insects), or plastic (reaching its highest perfection in man). The reader, however, must not be led to believe that the evolutionary trend has always been linear and direct. It has followed a tortuous process, with frequent dead ends; numerous lines have aborted or have not proved themselves capable of overcoming the passage of time. But the failures and the deviations of the transformist mechanisms have not arrested the increase in either anatomical or psychic complexity. This is the main point, for in the field of evolution, the final result is all that counts.

REGRESSIVE EVOLUTION

Certain biologists have emphasized the regressive character of evolution; uni- or multicellular organisms [those which contain chlorophyll and use solar energy for the synthesis of their own matter from simple chemical compounds, namely carbon dioxide (CO_2), water, and nitrogenous mineral salts (nitrogen contained in the atmosphere is rarely used)] are considered by these biologists to be superior to all others, because of their synthetic capacity. They depend upon no other living being for food supplies, while achlorophyllous beings (those lacking chlorophyll) do.* But are heterotrophic beings really regressed forms?

Autotrophic living beings, capable of their own synthesis and that of reserve substances, introduced into the aquatic environment a considerable amount of proteins, and sugars, which new beings, setting photosynthesis aside, did not fail to use.

These new animals, called heterotrophic beings, can metabolize the remains of autotrophic beings dissolved or scattered in the environment; they can also absorb, then digest, the autotrophic beings or penetrate them to obtain their nourishment (i.e., saprophitism, predation, parasitism). In all of these cases, the denatured and depolymerized macromolecules elaborated by the autotrophic beings are used by the

---

*Chemotrophic bacteria, still plentiful today, are considered the most primitive beings. Bacteria with chlorophyll (chlorobacteria) would be of more recent origin.

heterotrophic beings to carry out their own synthesis. These beings have, in effect, lost one faculty, photosynthesis, but have gained others which require many enzymes and long chains of chemical reactions.

To describe such evolution as regressive is tantamount to expressing an entirely subjective moral judgment. However, considering that new structures and abilities are acquired by heterotrophic beings, evolution could just as well be called progressive. These epithets reflect opinions which do not take into account the complete facts.

The loss of photosynthesis is commonly attributed to a mutation which eliminates chlorophasts (achlory) or acts against the synthesis of chlorophyll. Such a phenomenon, or one similar to it, occurred many times in some Protophyta: in the Euglenales, cryptomonadines, chrysomonadines, peridinians and Volvocales achlorophyllous and saprophytic species appear in varying numbers.

The various mutations affecting the chloroplasts of phanerogams at different times are still better known [see specifically Von Wettstein's (1957) study on the chlorophyllous mutants of barley]. An organism abruptly derived of its plastidial constituents will survive only if before mutation starts *it is provided with the enzymes* enabling it to digest and to restructure molecules of organic matter which then become its source of nitrogen and carbon.

Among the *Euglena,* whose feeding habits have been the object of intensive study, some species are strictly autotrophic; when deprived of light or CO_2 they soon die, even if their environment contains organic matter (*Euglena klebsii, E. pisciformis,* etc.). Other species are only optionally autotrophic; placed in the dark or in an environment deprived of CO_2 but containing compound organic matter (carbon + nitrogen), they multiply perfectly well (*Euglena gracilis*).

These *Euglena* are able to use the organic matter dissolved in the environment in which they swim. Numerous unicellular algae are capable of living either as autotrophic or as heterotrophic beings, depending on the conditions of their environment (for example, *Bracteacoccus* algae of the Lascaux caves); they are called mixotrophic animals. Potential mixotrophic capacity is the necessary condition for survival of achlorophyllous mutants.

Two conclusions must be drawn:

1. The shift from autotrophic to heterotrophic nutrition depends either upon a subtractive variation or upon the conditions of the environment.

2. In both cases, survival necessarily implies that the organism be equipped with an enzymatic "arsenal" and that its membranes have

specific properties (permeability) at the very moment when it changes from an autotrophic to a heterotrophic condition.

This "arsenal," described by some as preadaptive, can be explained by the occurrence of a "premonitory" mutation; too many enzymes are involved for it to be the result of a single variation. Although it is true that penicillinase, the enzyme which makes bacteria resistant to penicillin, is due to a mutation, one must keep in mind that such a variation involves only one and not several genes. It must also be remembered that the enzymes involved in the heterotrophic phenomenon have self-complementary effects on one another.

It would seem that the change from an autotrophic to a heterotrophic condition took place accidentally, as can be observed in modern *Euglena*; it was, however, preceded by an invisible evolution, which equipped the organism with a double "insurance" (auto/hetero) that was meant to be revealed by circumstances only.

A latent mixotrophic state is the *sine qua non* condition for the survival of achlorophyllous mutants.

How does Darwinism consider regression and the regressed form of the animal? Here, as elsewhere, the usual postulate—natural selection favoring the fittest—is put forward. Therefore, the loss of any organ, even of any function, in a species that survives is profitable to it. Actually, to admit that selection operates against the species would be nonsense. At the most, certain variations can be considered to be neutral. Broad mutations (or those considered as such) do not necessarily endanger animals. For instance, the loss of limbs in many lizards and in all snakes did not place these animals in a state of inferiority since they still appear in nature at present, whereas so many powerful lines of tetrapod reptiles have disappeared.

Snakes dwell in marine coastal waters, swamps, rivers, and forests. They comprise numerous arboreal (*Dendroaspis*) and underground (*Typhlops, Calabaria*) species. Apodal lizards live freely in meadows (*Anguis, Ophiodes,* etc.), take refuge in the sand (*Scincus*), or burrow underground galleries (*Feylinia,* amphisbaenians).

The transformation of a tetrapod reptile into an apodal reptile has occurred several times in distinct lines. This phenomenon can be observed in scincomorphs [such as Scincidae, Felidae, Cordylidae, Teiidae (*Scolecosaurus, Ophiomodon*)], among anguimorphs (such as Anguinidae), and in all the amphisbenians (except for the genus *Bipes,* which has retained its anterior limbs).

The essential condition for serpentine locomotion is the general elongation of the body. A squat animal cannot "undulate." Lengthening in the Squamata (lizards + snakes) is related to an increase in the number of presacral vertebrae; therefore, it is the trunk that lengthens; the sternum does not concomitantly increase in size. The ribs become supple and the body takes on a cylindrical shape. Moreover, the ribs and the segmental vessels increase in number as do the presacral vertebrae. The viscera—lungs, kidneys, blood vessels, and urinary and genital ducts—participate in the stretching of the body. The relative position of the kidneys and testes or ovaries is modified longitudinally.

In boid, anilid, and hydrophid snakes, two lungs exist and function, but the right one is three times as large as the left, which is rudimentary in several grass snakes and the Crotalidae and totally missing in the Viperidae and Hydrophidae (marine snakes).

The elongation of the right lung is so great in some species (*Enhydrina*, *Acrochordus*, etc.) that this organ extends as far back as the cloaca! Certain parts of the stretched lung do not perform a respiratory function; only the anterior part (first third) shows an alveolar structure, the other two-thirds consisting of an air sac whose function is still unclear.

In some snakes the wall of the trachea may form a membranous evagination (*Naja*) or take on a more or less alveolar lung structure (Viperidae). The specific attachment of the mandible to the skull allows a considerable enlargement of the buccal opening and consequently the absorption of bulky prey. Because of the strong evidence available, we can assert that the entire anatomy of ophidians was altered.

Although the evolution of the apodal Squamata has in some respects been suppressive, it cannot, however, be considered regressive; it has produced forms with a perfectly balanced anatomy and physiology and that are also extremely vigorous and resistant to the variations of the environment. They have been provided with an extremely elaborate weapon: venoms, which result from a most extraordinary chemical innovation.

The evolution of snakes did not occur in a disorderly manner. It was directed and preceded by a thorough change in the general body form. Species with an elongated and cylindrical body became apodal or subapodal. If, as the Darwinians claim, the loss of limbs had occurred by subtractive mutations, it could only have occurred subsequent to the elongation of the body when locomotion became undulating and the two pairs of legs, too far apart, no longer sustained the trunk high enough above the ground.

In various lizards the legs are rudimentary; in some rare species they perform limited functions, such as in the Mediterranean seps (*Chalcides lineatus*) which uses them for slow walking but for rapid motion folds them against its body into depressions ("coaptations") and undulates in the manner of a snake.

In fact, the one variation that determines all the others is the numerical increase of the presacral vertebrae. This is proof that the *evolutionary "value" of any variation depends upon its timely appearance.*

There is a relationship between the number of presacral vertebrae and the proportions of the body. Here are a few figures from Guibé (1970) concerning lizards: *Eumeces schneideri:* 30 presacral vertebrae, normal anterior and posterior limbs; *Lygosoma punctulatum:* 36 presacral vertebrae, reduced anterior and posterior limbs, loss of a phalanx on toe IV; *Chalcides lineatus:* 48 presacral vertebrae, reduced limbs, loss of all phalanges on toe I; *Seps tridactylus:* 52 presacral vertebrae, reduced limbs, total loss of toe I, loss of the phalanges on toe V; *Anguis fragilis* (blind worm): 65 presacral vertebrae, anterior and posterior limbs absent. The number of presacral vertebrae exceeds one hundred in the amphisbenians and in *Dibamus* [which according to Underwood (1957) is an aberrant gekko and not a skincomorph], all apodal and hypogeal animals.*

The "prime motive" of the snakelike structure lies in a change in embryogenesis; the formation of new vertebrae requires additional new somites; the number of segmental nerves and arteries becomes equal to the number of vertebrae (see Raynaud, 1963). As is often the case, the variation first concerns embryogenesis, and the correlations observed after completion of ontogenesis are necessarily preceded and prepared by chains of chemical reactions appearing in a determined order.

Raynaud (1972) has recently resumed his study of the embryology of apodal or vestigial-limbed reptiles. His observations and experiments indicate that, as noted in birds, the mesodermal metameric masses, or somites, are the elements which induce and organize the limbs. In the Apoda, the somites apparently lost this property. Raynaud confirmed that from the beginning the somites of the embryos of apodans are much more numerous than those of tetrapods: 32 in green lizards, 70 in blind worms, 173 in grass snakes.

The total or partial interruption of the growth of limbs can be explained by the loss of the inducing ability of the somites in the trunk region, but we do not have the slightest idea of what can cause the number of somites to increase and thus cause the corresponding thorough change in the structure of numerous parts of the embryo.

---

*We are not listing these species in phyletic order. The apodal condition has occurred in eleven families of lizards and in all amphisbaenians and snakes, of which some species have vestigial limbs (*Bipes* in the former, and *Boa* in the latter).

The layout of the muscular system clearly indicates the orderly manner which serpentiform reptiles are formed. The undulating locomotion (creeping) is only possible with a skeletal, muscular, and nervous apparatus repeating itself along the vertebral axis and a means by which the muscles contract in rhythmic sequence controlled by the appropriate nerve impulses. The great development of intersegmentary aponeuroses on which muscles are attached is one of the peculiarities of the musculature; as the bony surfaces are highly reduced, they could not anchor the powerful "reptatory" muscles.

Thus, from all points of view, the musculature of the episoma of serpentiforms is very specific and suited to the mode of locomotion adopted by the animal. The different types of musculature in the Boidae, Viperidae, and Colubridae correspond to different types of "reptation" (creeping locomotion).

The prime importance of this correlation is that it is proportional to the number of extra presacral vertebrae. *Lygosoma*, which has only 36 presacral vertebrae, exhibits a limited reduction of its limbs, losing only one phalanx on toe IV.

Snakes do not show those brutal mutations which pertain to the pathology of genes and thus belong to teratology. The result of the loss of limbs is the production of a harmonious being (in spite of its aberrant appearance) whose different parts are strictly correlated. Snakes and blind snakes are not monsters or failures resulting from a series of random events.

The reduction of limbs is in no way aleatory since it is determined by the number of presacral vertebrae. We know that these arguments, no matter how factual and reliable they are, will not convince doctrinaires. We know what their answer is likely to be: The rate of mutation is so high that the animal must find the very one it requires. But assumptions remain assumptions, and no one has yet found in the broods of four-legged lizards a great number of monstrosities, among which would appear the mutation rendering the reptile both apodal and perfectly adapted to a new way of locomotion and life.

This survey (although it does not go into the detail of the structures) shows that, if one admits that mutations gave rise to serpentiform reptiles, an infinity of mutations had to take place to produce the specific mutation that was "tailored" to the preexisting structure and that enabled evolution to advance.

The "random" explanation comes up against a great many other obstacles. Serpentiform evolution occurred in several independent groups (at the family level or below) which, except for the Amphisbaenidae and Ophidae, comprise tetrapodal stout types which are lacertiform to greater or lesser extent, types with reduced limbs, and also apodal types. In other words, the same evolutionary trend took place at the expense of fairly different genetic pools, and also rapidly, since in three groups (Scincidae, Cordylidae, and Anguidae), the original species and the serpentiform species still exist.

The history of these three lines is not well known. The oldest Scincidae have been found in the Upper Pliocene, but we have no idea of the stage of development of their limbs. The Cordylidae appear toward the end of the Jurassic and the Anguidae in the Cretaceous. It is likely that they are more ancient than the rocks in which the presently known fossils were contained.

The evolution of the present saurians is very slow, not to say nonexistent. In most groups for which fossil evidence has been found, we are certain that since the Miocene they have only undergone very slight variation. It seems that tetrapodal and apodal species are equally stabilized; the serpentiform apodal structures have become so highly specialized that they can progress no further. Like all living beings, they are likely to undergo somatic mutations while retaining their same structural plan and way of life.

Despite their structure, serpentiform reptiles show an adaptability for living in very different environments (terrestrial, aquatic, subterranean, arboreal); thus, loss of limbs, when compensated by undulating locomotion, can lead to numerous adaptations.

THE LIMITS OF EVOLUTION

Evolution tends to make more complex both structure and function, but by increasing the animal's level of "psychism" it gradually liberates it from the ascendency of the environment. Those who consider the loss of certain functions as a process contrary to progressive evolution are making a mistake. They are only taking into account one aspect of the history of animals as it unfolds on the earth's giant stage.

Has nature lacked imagination in its creative effort? This is possible, but before resorting to science fiction, one must stop and think. For instance: Why hasn't evolution fostered half-vegetable, half-animal beings? One can easily imagine a worm bearing on his back a long chain of transparent cells containing chloroplasts producing sugars and proteins from very simple substances: water, CO_2, nitrates, oxygen, etc.

A fish, equipped with filamentous crests filled with chloroplasts and capable of absorbing nitrates dissolved in the upper layers of the waters where light penetrates, would not need a digestive tract. Although mobile and sensitive, it would feed like algae. But wouldn't this theoretically conceivable being incorporate two irreconcilable characteristics? Motility requires a considerable amount of *readily available* energy while photosynthesis, in order to provide any sizeable organism with nourishment, requires that large surfaces be exposed directly to a source of light: hence the stretching and flattening of the thallus of algae and the great number of families of phanerogams. Such a chlorophyllous apparatus, because of its bulk and cumbersomeness, would make rapid and coordinated movements awkward and even impossible.

Moreover, the absorption of mineral salts, occurring during photosynthesis, should theoretically be performed by the surface of the body immersed in the water containing the dissolved salts. Plants became terrestrial by developing an absorption system for extracting water and salts from the soil, but they consequently found themselves riveted forever to the ground.

Thus, the only beings both motile and capable of photosynthesis (because of their chloroplasts) are aquatic microorganisms; one-celled organisms (Protophyta) such as *Euglena*, Cryptomonadiniae, peridinians, or, in a few cases, several-celled bodies (Volvocales, such as *Pandorina*, *Volvox*).

The symbiosis between Cnidaria (Hydraria, Actinaria) and some colored peredinians is less important and sets the limits to what can result from the association of photosynthesis and motility. These cnidarians are immobile or show very little motility. Can one say that the first aquatic uni- or multicellular beings regressed when they lost their chloroplasts and modified their metabolism? Doubtfully so, since at the same time they acquired innumerable potentialities and freed themselves from the heavy burden of immobility.

Evolution, when beings capable of feeding at the expense of other beings appeared, did not take a step backward but followed a path that proved to be much more fruitful than the first with regard to evolutionary potential.

II

Creative Evolution, or the Appearance of Types of Organization

GENERAL REMARKS

The formation of the phyla or basic structural plans constitutes the most important and, perhaps, the essential part of evolution. Each phylum offers great novelties and its structural plan guides the destiny of the secondary lines.

Since paleontology does not shed any light on the genesis of the phyla, one must have recourse to the data drawn from comparative anatomy and embryology. These sciences, their value notwithstanding, do not enable us to reconstruct the past with a high degree of certainty; the simulation of evolution is therefore a task which, up to now, was beyond the means of the biologist.

According to Haeckel's (1866)* fundamental biogenetic law, embryology gives us some idea of what the ancestors of present invertebrates could have been like; we cannot, however, expect too much from it since, in a great many cases, ontogenesis is deeply modified by cenogeneses and adaptations to very particular modes of life: We are referring to pelagic or planktonic larvae with supposedly archaic characters which may not be so archaic after all.

Evolution must be studied at a low systematic level, not above that of the genesis of classes or orders, since for some of them we have suffi-

---

*Haeckel mentioned the fundamental biogenetic law in several of his writings. In his "Generelle Morphologie" (Haeckel, 1866, Vol. II, p. 300), he sets forth for the first time: "Ontogeny is a repetition, a short and rapid recapitulation of phylogeny, following the laws of heredity and adaptation."

cient fossil evidence. Prior to setting forth our views on the mechanisms and course of evolution, we shall clarify certain points regarding structural plans and the distribution of characters among the various levels of the systematic hierarchy, going from the general to the specific.

A phylum is characterized by its structural plan and by specific chemical compounds which indicate specific fundamental metabolic characteristics. A *class* is elaborated by adding characteristics to the fundamental plan, without reaching specialization, however; it is at the level of an *order* that specialization and idiomorphon (morphological type) appear freely. Thus, to the fundamental plan are added some characteristics (those of the class) which are slightly more specific than those of the higher category, and finally the specializing characteristics of the idiomorphon appear.

These systematic groups can be classified in many different ways; we will consider the following two:

1. The classic one, in which categories are superimposed in a diagram like a genealogical tree, since it is true that the systematic sequence is also the chronological order of appearance (Fig. 2).

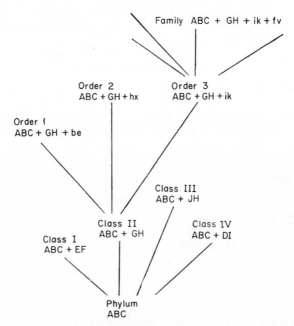

Fig. 2 Diagram showing the composition of a phylum, i.e., the systematic unit corresponding to a major structural type of organization. ABC, characters specific to the phylum; EF, GH, JN, DI, characters specific to the classes; be, hx, ik, characters specific to the orders; fa, characters specific to the family.

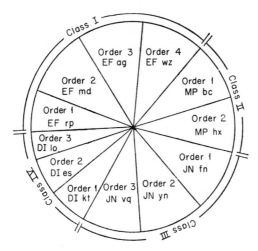

Fig. 3 Representation of the composition of a phylum by comprehensive groups; in capital letters, the characters specific to the classes; in lower-case letters, the characters specific to the orders.

2. The comprehensive one (Fig. 3), the most comprehensive being the phylum, which includes them all. In this diagram, the phylum appears as the entire circle, divided into as many sectors as there are orders in the phylum, the surface of the orders being proportional to the number of species included in it. This subdivision can go down to the species level. This diagram has the advantage of showing at a glance the relative importance of each systematic group. In Fig. 3, we have not gone further down than the level of the order.

The animal kingdom consists of approximately 21 phyla: Protozoa, Porifera, Cnidaria, Ctenophora, Platyhelminthes, Nemathelminthes, Rotifera, Annelida (including, apart from the so-called annelids, the Sipunculida, Echiurida, Priapulida), Endoprocta, Phoronida, Bryozoa, Brachiopoda, Chaetognatha, Pogonophora, Mollusca, Onychophora, Tardigrada, Arthropoda, Echinodermata, Protochordata, and Vertebrata (see Appendix). Some comprise thousands of species, others only a few dozen.

The genesis of the fundamental structural plans of these phyla was the greatest achievement of evolution, and it is likely that the genesis of the classes was achieved through the same mechanisms.

As noted earlier, paleontological evidence concerning the origin of the phyla is poor. Since the animals concerned are very ancient, most of their fossils have been destroyed by subsequent metamorphosis of the rock in which they were embedded. In an attempt to illustrate this fact, we shall sum up the little we know about the initial history of the Arthropoda, the richest phylum with regard to patterns and species.

As early as the lower Cambrian, this phylum was represented by fossils belonging to groups already well characterized: the Trilobita, Merostomata including *Paleomerus* (with doubtful affinities, found in the lower Cambrian of Sweden), pseudo-Crustacea, and phyllocarid Crustacea (*Dioxycaris, Osoxys*). Their Archean predecessors are still unknown; however, *Parvancorina*, a large fossil with a segmented body belonging to the Ediacara Precambrian fauna, has been interpreted by some paleontologists (Termier and Termier, 1968) as a possible giant larva of an arthropod (possibly related to the protaspis larva of the Trilobitomorpha). If, however, one agrees that the Arthropoda were derived from some primitive annelid stock (which is very likely), then *Parvancorina* could also be a "prearthropod" showing very archaic annelid characteristics.

We are in the dark concerning the origin of insects. Their most ancient fossils belong to the Collembola (*Rhyniella*) and have been discovered in middle Devonian sediments (Old Red Sandstone). They are closely related to the present Podura. Massoud (1967) includes *Rhyniella* in the family, Neanuridae which has living representatives and already shows a high degree of evolution; one can thus say that the genesis of the Collembola occurred prior to the middle Devonian. Massoud's conclusion is that "the Collembola have evolved very little" since that remote period (380 million years ago).

A similar situation can be found in the Chelicerata. The earlier forms of scorpions (Protoscorpionidae) date back to the upper Silurian (Siluronian) and seem to have been aquatic; they are already fully characterized, but the first abdominal segment persists in the adult and the preabdomen is composed of eight tergites.

As suggested by Störmer (1955), an authority on paleozoic arthropods, *Paleomerus hamiltoni* and its analogues (*sensu lato*), the species of *Aglaspis*, all from the Cambrian, are the likely ancestors of the Merostomata. *Paleomerus* (9 cm long) had a well-defined head, followed by 11 or 12 segments which form the opisthosoma, and a telson. From all points of view it is a typical arthropod, resulting from a lengthy evolutionary process of which no links have yet been found.

In fact, all Cambrian arthropods, except the Trilobita, which are totally extinct, easily fall into modern classes. It is to be hoped that the Precambrian sediments, namely at Ediacara, will provide lucky findings among which some forbearers of the Arthropoda will be found. *Parvancorina* could be one of these ancestors.

From the almost total absence of fossil evidence relative to the origin of the phyla, it follows that any explanation of the mechanism in the creative evolution of the fundamental structural plans is heavily burdened with hypotheses. This should appear as an epigraph to every book on evolution. The lack of direct evidence leads to the formulation of pure conjectures as to the genesis of the phyla;we do not even have a basis to determine the extent to which these opinions are correct.

Classes, from what we know of them (refer in particular to the transformation of reptiles into mammals) arise from variations of unequal scope and, above all, from the acquisition of characteristics which gradually build upon the fundamental plan. None of the "spectacular" anatomical disruptions which had been predicted by certain biologists have been observed. The "systematic mutations" conceived by Goldschmidt (1944), that were meant to transform radically the structural plans and give rise to new phyla, have never appeared in the history of the amphibians, reptiles, and mammals. They are pure fantasy. The youngest classes of all, mammals and birds, date, respectively, from the upper Triassic (approximately 200 million years ago) and from the Jurassic (approximately 135 million years ago).

THE APPEARANCE OF A MAJOR PATTERN OF ORGANIZATION: THE MAMMALS

Although we have no knowledge of the origin of the branching points, fortunately a number of fossils are available that provide us with a fairly precise idea of the formation of a few major groups of animals. For instance, the genesis of mammals from reptiles is rather well known.*

In the last few years, views on the history and composition of the class Reptilia have changed a great deal. The hypothesis that this class was dichotomous from the very onset has been successfully disproved by what seems to be reliable evidence. In paleontology, however, the discovery of a new fossil can considerably modify our views and make interpretations obsolete which were previously thought to be definitive.

*Those readers who are not zoologists will probably find the pages describing the history of reptiles lengthy and tedious. We prefer to take that chance rather than to take the easy way out. We wish to use facts to show the actual rise of a new major pattern of organization. It is the best way to bring to light the discrepancy between reality and theory.

FROM CAPTORHINOMORPHS TO PELYCOSAURS*

All reptiles (fossil or living) are believed to have been derived from a single common stock: the Captorhinomorpha, which inhabited the emergent lands of the Northern hemisphere during the long period extending from the middle Carboniferous to the middle Permian. The skull showed no temporal fossae; the massive stapes extended from the fenestra ovalis to a long depression in the quadrate where it articulated. All the known parts of the anatomy are archaic and unspecialized. The habitus was similar to that of bulky lizards, with short limbs and with the humerus and the femur held horizontally; several of their features were similar to those of their amphibian ancestors, the stegocephalian labyrinthodonts. Among the Captorhinomorpha are placed the Diadectomorpha, considered by Watson (1954) as the original stock of the Sauropsida (several lines of reptiles and birds); recent work has led to the suggestion that diadectomorphs could even be true amphibians!

The Captorhinomorpha radiated into several lines whose relationships and affinities are difficult to define clearly. We will only concern ourselves with the Synapsida, characterized by the presence of a single temporal fenestra in each side of the skull that is very similar to the fenestra of the Captorhinomorpha and remains archaic in structure.

The order Pelycosauria, or Theromorpha, is placed at the base of the synapsid phylogenetic tree; pelycosaurs appeared during the upper Carboniferous, flourished during the lower Permian, began declining from the middle Permian, and became extinct during the lower Triassic. Their most ancient and archaic species, close to the Captorhinomorpha, are grouped in the suborder Ophiacodontia, which is divided into two families. The less evolved is the Ophiacodontidae, whose members, because of their long snout and identical seizing teeth, paleontologists consider to have been piscivores; the members of the second family, the Eothyrididae, were clearly carnivorous. The two other anatomically specialized pelycosaurian suborders included both carnivorous species, the Sphenacodontia, equipped with a high "dorsal sail" (a large dorsal fin stretched across the spiny vertebral apophyses), and herbivorous species, the Edaphosauria, with very numerous rounded palatine teeth and a lower jaw bearing identical teeth implanted on a lateral internal extension facing the palatines, the whole complex forming a powerful grinding mill. Their most extreme genus is *Cotylorhynchus*, of the lower Permian. The Ophiacodontia, and possibly the Varanopsidae among the Sphenacodontia, seem to be the only Pelycosaurs which, being fairly un-

*See the Classification of Reptiles, p. 44.

specialized (without dorsal fin), might have given rise to the Therapsida.

The mammalian or premammalian characters of the Pelycosauria are very few and do not stand out, but they do exist (see p. 44).

THERIODONTS: MAMMALIAN ANCESTORS

From the middle Permian, pelycosaurs were supplanted by the therapsids which, in turn, disappeared during the upper Triassic; they survived in the form of mammals, their natural heirs.

The history of the order Therapsida* is of paramount interest: it reveals how a class with many novelties—the mammals—arose. Its fossil evidence is extensive and often well preserved. The Therapsida is divided into three suborders: the Phtinosuchia, Theriodontia, and Anomodontia.†

The most ancient and the least evolved are the phtinosuchians.‡ They looked like a common lizard, with a large and high head, a narrow snout, eyes very far back, and a pineal opening. The high temporal fossa slanted toward the back. The skull articulated with the vertebral column by a single condyle. The orbit was separated from the very posterior temporal fossa by a thin temporal arch formed by the postorbital and the jugal. The jaws bore strong teeth which differentiated into narrow incisors, much longer canines, and numerous high and sharp postcanines. This seizing and lacerating dentition is typical of carnivorous animals. These phtinosuchians were large sized; the skull of *Phtinosuchus* was 20 cm long. They have only been found in the upper Permian of Russia.

With the theriodonts, which seem to be descended from the phtinosuchians, the approach toward the mammalian type becomes increasingly clear (Fig. 4). Their numerous genera can be divided into three main lines: the Gorgonopsia, the Cynodontia, and the Therocephalia. The latter two can be further divided into several secondary groups.

The gorgonopsians, by their high and long skull with two postfrontals, resemble phtinosuchians. The reduced temporal fossae, which are not in contact with each other, are similar to those of pelycosaurs. In

*Since mammals make up a class, therapsids should logically be placed above them, although the latter, when considered within the complex of reptiles, only constitute an order. The evolutionary approach blurs the systematic hierarchy.

†We will not concern ourselves with anomodonts or titanosuchians (classified close to gorgonopsians), which have remained alien to the history of mammals.

‡Phtinosuchians were formally classified among theriodonts, within the infraorder Gorgonopsia.

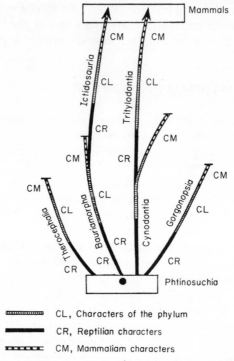

CL, Characters of the phylum

CR, Reptilian characters

CM, Mammalian characters

Fig. 4 Diagram representing the distribution of the reptilian and mammalian characters, as well as those specific to the lineage, in the various "twigs" of theriodonts. CL, characters of the lineage; RC, reptilian characters; MC, mammalian characters.

sum, the first gorgonopsians were smaller than pelycosaurs, their predecessors, but not unlike them. Although retaining some primitive characters, the gorgonopsians acquired mammalian characters, but fell into excessive predatory specialization. The huge canines (*Digorgodon*, for instance) are similar to those of the Quaternary *Machairodus*, while the teeth posterior to the canines are so limited in number and in size that they are of no use in the mastication process. These predators grew to large sizes (the skull of *Digorgodon* reached nearly 37 cm); the cranial bones became very thick and sculptured.

The Gorgonopsia, although not in the direct line of descent of the mammals, possess unquestionable mammalian characteristics. The enlarged dentary definitely foreshadows the formation of a coronoid apophysis (coronoid process). The jaw muscles are partly attached to it; the reduced quadrate no longer closely interlocks with the squamosal, which is equally reduced; this is the beginning of the release of the quadrate which, when completed, will form the incus. The stapes which, in pelycosaurs, was a massive bar extending from the fenestra ovalis to the quadrate and to the paroccipital, grows very much lighter and tends to adopt a plate-like structure pierced by the stapedial fenestra, through which passes the stapedial artery. The fairly slender limbs were probably more erect than those of captorhinomorphs and phtinosuchians. The gorgonopsians were a short-lived group. They disappeared for unknown reasons (perhaps because of their specialization) at the end of the Permian.

The theriodonts, gorgonopsians excepted, form a complex of species showing a greater or lesser number of mammalian characteristics. They have been divided into five suborders: the Cynodontia, Tritylodontia, Therocephalia, Bauriamorpha, and Ictidosauria. Their respective characteristics are rather easily identifiable, but their phylogenetic relationships are difficult to define.

Paleontologists agree on the existence of an evolutionary trend going from the primitive cynodonts of the upper Permian to the tritylodonts and Haramyidae. The most interesting ictidosaur suborder has been related by some zoologists (Hopson and Crompton, 1969) to the Cynodontia, and by others (Ginsburg, 1970) to the Therocephalia–Bauriamorpha.

All paleontologists (e.g., see Crompton, 1963) note that the acquisition of mammalian characteristics has not been the privilege of one particular order, but of *all the orders of theriodonts*, although to varying degrees. "This progressive evolution toward mammals has been most clearly noted in three groups of carnivorous therapsids: the Therocephalia, Bauriamorpha and Cynodontia each of which at one time or another has been considered ancestral to some or all mammals."

The shaping of the mammalian form, which lasted from 50 to 60 million years, occurred in a smooth and gradual manner. The evolutionary tendencies present in *all* the theriodont lines are: in the mandible, the upper end of the dentary extends posteriorly into a flat process which, in mammals, is called the coronoid process, while the lower end is prolonged horizontally. The postdentary bones (angular, articular, prearticular, supraangular) undergo a strong reduction.

Homeothermy, a feature that does not belong to mammals exclusively, seems to have been present in a great many therapsids, if not in all. The bones are fibrolamellar, highly vascular, and very similar to the Haversian bones of mammals and birds. This structure is correlated with continuous growth which was made possible by constant body temperature. This structure developed in theriodonts, among which its evolution was long and gradual. It has been suggested that the homeothermy of the most evolved species was similar to the imperfect homeothermy of modern monotremes.

In reptiles with a variable body temperature (poikilothermic), bone has a discontinuous growth; it shows a zonal lamellar structure which is formed in successive concentric layers, corresponding presumably to the succession of seasons (de Ricqlès, 1972a,b).

Before they had attained the premammalian stage, theriodonts were no longer cold-blooded animals; they were less dependent on the conditions of the environment and less sensitive to the wide variations of the external temperature. Although we do not clearly know why, homeothermy may have been the preliminary condition to the "mammalization" of theriodonts. A. Ricqlès has listed the consequences of homeothermy on the animal's physiology and behavior: higher general metabolism and intense phosphocalcic metabilism, quicker and continuous growth, and greater activity.

THE CYNODONT LINE

The most ancient Cynodontia (upper Permian) show characters relating them to the scaloposaurid bauriamorphs, but their structure is increasingly "mammallike": the occipital condyle is clearly divided; the epipterygoid is fused with the prootic bone; in the mandible, the angular has only a reduced, reflected lamina. Cynodonts are the only group to possess a prootic with a protruding ridge. The secondary palate is complete; the cheek teeth, although small and numerous, bear several cusps. Both features are also present in advanced scaloposaurid bauriamorphs; moreover, the occlusion between upper and lower cheek teeth is incomplete in both groups. In cynodonts, palatosuborbital spaces are lost; this archaic feature was preserved in a few members of the Procynosuchidae (a family of Permian cynodonts) and lost in Triassic cynodonts.

The pattern of the occlusal face of a number of cheek teeth, including cuspid *g* (Kühnecone) of *Thrinaxodon*, resembles that of *Diarthrognathus* (= *Eozostrodon?*). According to Crompton and Jenkins (1968), this similarity is evidence of the derivation of *Diarthrognathus* and allied genera from cynodonts of the *Thrinaxodon* type. However, thrinaxodontid cynodonts possess a great number of reptilian characters such as alternate replacement of teeth, imperfect occlusion between the upper and lower cheek teeth, and presence of expanded ribs fused with the lumbar vertebrae.

One must keep in mind that cynodonts, apart from the Thrinaxodontidae, are anatomically too specialized, either in the carnivorous direction (*Cynognathus*, for example), or in the herbivorous direction

(*Diademodon,* for example) to belong to the direct ancestry of any mammalian group. In any case, this is Hopson and Crompton's opinion, which we think to be well founded. These cynodonts have, however, undergone an evolutionary process towards "mammalization"; the life history of the Traversodontidae exemplifies this phenomenon. The Traversodontidae were herbivorous cynodonts that lived in the middle Triassic in southern Africa and South America. The most ancient types, such as *Scalenodon,* had cheek teeth similar to those of *Diademodon,* which might indicate a possible ancestry. In the Traversodontidae, these teeth evolved and became capable of cutting and grinding; simultaneously, the articulation of the jaw acquired a greater freedom, which allowed movements from front to back as well as transversely; the canines became smaller in size and some incisors became longer and took a very slanting position.

Paleontologists admit that, from the upper Triassic to the middle Jurassic, the traversodontid Cynodontia evolved into tritylodonts.

"PREMAMMALIAN" REPTILES DERIVED FROM THE CYNODONTS

Although specialized, tritylodonts* have obvious affinities with mammals. Their orbit, which is not closed at the back by a postorbital bar, is confluent with the temporal fossa. The prefrontal, postfrontal, and postorbital are lost, as in cynodonts; the base of the skull resembles that of cynodonts. There is no pineal opening in the cranial vault. The teeth are differentiated into incisors, the second of which is caninelike (six on each jaw), and into cheek teeth similar to those of the multituberculate mammals which appeared in the Purbeckian (upper Jurassic) and became extinct in the Eocene. A long diastema separates the canines from the cheek teeth, which bear longitudinal lines of crescent-shaped cusps (three on the upper jaw, two on the lower jaw). Teeth, as in mammals, are deeply rooted into alveoli; the upper cheek teeth bear two anterior and three posterior roots, the lower cheek teeth only have two, one anterior and one posterior.

The lower jaw is of major interest to us. It is formed almost entirely by the dentary; it is of the mammalian shape with an ascending ramus extending into an apparently well-developed apophysis (the ascending

---

*Substantial remains of the skeleton of three genera have been found: *Tritylodon* (upper Triassic) of South Africa, *Bienotherium* (upper Triassic) of China, and *Oligokyphus* (upper Rhaetic, Charmouthian) of Europe.

ramus is incomplete in the available specimens). The quadrate is small (4 mm high) and loosely attached to the skull wall by ligaments. Its suspension is of the streptostylic type. The articulation of the lower jaw with the skull was still between the quadrate and the articulate; it was rather weak, but it only partially bore the action of the jaw muscles which, since they were attached to the dentary at the level of the teeth, relieved the strain on the articulation.

The squamosal is a relatively fragile bone against which the external auditory duct is in part molded; it forms a glenoid cavity which is not functional since no jaw bone comes into contact with it. Devillers (1961b, p. 204) writes: "the location of the presumptive cavity is already visible although the articulation of the dentary is not yet formed." We shall come back to this disconcerting fact.

A great many other characters show the tritylodonts to be "premammals": they possess a bony palate which extends posteriorly almost as far as the dental arcades; a jaw articulation which is located high above the cheek teeth; and two occipital condyles. The strong zygomatic arches, the sagittal and lambdoid crests which raise the skull surface, and the wide temporal fossa, which is confluent with the orbit, are also typical mammalian features. The prefrontal, postfrontal, and postorbital have disappeared.

Tritylodonts retain, however, many anatomical features which are typically reptilian. In the skull, the parasphenoid is well developed, whereas it disappears in mammals (becoming vestigial in a few species); the temporal fenestra is still present. The brain cavity is small, relatively much weaker than in mammals, even in those with the least evolved brains (monotremes, marsupials).

The vertebral column is not divided into distinct regions; all presacral vertebrae bear ribs. The posture of the limbs is still transversal, but the cubitus (ulna) has a form which seems to allow limited rotation of the radius over it, announcing the pronation and supination that most mammals are capable of performing.

The girdles, still partially unknown, are predominantly reptilian. In the pectoral girdle, the coracoid and the scapula form the glenoid cavity; the reduced precoracoid is fused with the scapula. The structure of the pelvic girdle, despite Kühne's (1956) and Devillers' (1961a) opinions, is not clearly mammalian.

Consequently, the genera *Tritylodon, Bienotherium,* and *Oligokyphus,* although very closely approaching mammalian structure, still belong to the class Reptilia. One must necessarily be very cautious when assigning a systematic position to these genera, and all the more so, as we do not know whether these animals possessed a scaled or hairy covering, or whether their body temperature was constant or fluctuated with that of the external environment.

If one wants to judge with accuracy the evolutionary value of the tritylodonts, one must keep in mind that these animals probably had a very specific way of life and fed on plant food requiring a powerful grinding and grating action, which their teeth seem to have been capable of performing. The significance of the protrusion of their incisors, which do not have the characteristic chisellike form of rodents, is yet to be discovered. The incisors of *Bienotherium* and *Oligokyphus* are somewhat reminiscent of those of the insectivorous Soricidae, which catch small live prey in a clawlike movement; but this is only a rough guess. One can only say that with such a peculiar dental anatomy, tritylodonts must have had specialized habits. Thus, they form a branch of theriodonts which gave rise to no later forms—perhaps because of their excessive specialization—but which closely approached mammalian structure.*

Another cynodont line provides evolutionists with highly interesting facts. We are referring to the South American family Chiniquodontidae, which Romer (1970) described in a masterly fashion. This family lived in the middle Triassic. Its most remarkable feature is the presence of an incipient squamosal–dentary articulation, attained in *Probainognathus jenseni.*

The dentary, with a well-defined lower posterior angle, bears a very strong coronoid process; the posterior tip of this process extends backward in the form of a triangular lamina which is applied to the dorsolateral surface of the supraangular, terminating just short of the posterior articular expansion of the latter bone. The terminus or "spur" of the dentary, together with the distal tip of the surangular, is almost in contact with the squamosal. Now, most important for evolutionists is the fact that the squamosal opposite this "spur" of the dentary shows a shallow "cup" which seems to initiate the formation of a glenoid fossa. One can thus observe the stage immediately preceding the dentary-squamosal articulation, which is the utmost characteristic of mammalian structure.

Romer notes that chiniquodontids (*Chiniquodon, Belesodon, Probelesodon, Probainognathus*), all from the middle Triassic, have an unspecialized structure, and are thus free to evolve in various directions. They may have been derived from the Galesauridae (*Thrinaxodon*), and they approach mammals by their possession of a fully developed secondary

---

*It was previously thought that tritylodonts and multituberculates belonged to the same mammalian subclass; Watson's work has shown this opinion to be wrong. Resemblances between tritylodonts and multituberculates would merely be superficial.

palate and of an almost perfected dentary-squamosal articulation. However, their dentary does not yet bear an articular condyle; its "spur" is still flat, with a pointed end; and their quadrate is remarkably weak.

Signs indicating a dentary–squamosal articulation have been reported in *Trirhachodon* and *Massetognathus* (both traversodonts) in which there is a distinctive hump on the ventral edge of the squamosal in the region of the jaw articulation. Crompton (1963) and Romer (1970) believe that this hump reinforced the articular–quadrate jaw joint, which, because of the small size of the articular, is rather feeble.

In fact, as will be more clearly seen in ictidosaurs, since the discovery of intermediate types the systematic reptile–mammal contrast based upon the fundamental anatomical criterion of the mode of articulation of the jaw is no longer so clear-cut. Moreover, the lower Triassic, Jurassic, or even Cretaceous mammals retain traces of the reptilian pattern in their jaw structure.

In any event, taking into account their mammalian features and their structure of a generalized type, Romer considers that the chiniquodontids are the theriodonts which best combine the necessary conditions to be the direct antecedents of mammals.

THE THEROCEPHALIA–BAURIAMORPHA

The second major line of theriodonts was begun with the Therocephalia which may have been derived from an archaic gorgonopsian, or even from a phtinosuchian; they attained very early the therocephalian type which gradually became the bauriamorph type.

Their anatomical characters are archaic. The massive skull, with a long and flat snout, shows two temporal fossae very close to each other; this characteristic distinguishes them from gorgonopsians. The epipterigoid is well developed. There is a clear, although discrete, tendency toward the reduction of the number of bones; the parietal is missing, as are the postfrontals in many cases. The secondary palate is absent, but in a few specimens (*Whaitsia*, for example) the concave roof of the buccal cavity somewhat foreshadows the partition by a secondary palate.* The phalangeal formula is mammalian (2,3,3,3,3), not reptilian. The quadrate and the quadratojugal are small in size. The Therocephalia did not

---

*Evidence has recently been produced indicating that the most primitive therocephalians (the Pristerognathidae) already exhibited an incipient secondary bony palate and two canines on each side of the upper jaw, thus showing a clear tendancy to acquire mammalian status (Mendrez, 1972).

escape specialization and have exaggerated their features of predaceous carnivores: large incisors and canines, and sharp cheek teeth, which are lost in the family Whaitsiidae. Some of them were fairly large (skull measuring approximately 30 cm).

Like gorgonopsians, therocephalians became extinct at the end of the Permian. In fact, they became bauriamorphs so gradually that a number of paleontologists place both groups in a single systematic category.

Bauriamorphs seem to have evolved from the Pristerognathidae, which are placed at the base of the Therocephalia line. They are presently believed to have evolved into two main lines, the first comprising the Ictidosuchidae and Bauriidae, the second the Scaloposauridae and Eriolacertidae. The Ictidosuchidae include archaic forms that are close to the Therocephalia and lack a secondary palate; *Ictidosuchops*, the best-known genus, lived from the upper Permian to the beginning of the Triassic in South Africa (upper Tatarian). The Bauriidae show a clear advance over the preceding family: A secondary palate is present but the palatines do not participate in its formation; the occipital condyle tends to become double, and the pineal opening is lost. The skull roof is drawn out in a fairly high sagittal crest. The postorbital bar is interrupted.

The postcanine teeth (= cheek teeth) are of interest to us because their occlusal face bears small cusps. The structure of the lower jaw is reptilian: The postdentary bones are highly developed, especially the articular, but the dentary is nonetheless predominant. Thus, the mammalian characteristics of the Bauriidae are the following: the narrowness of the facial block anterior to the protruding zygomatic arches, the very anterior position of the nares (a rather unimportant characteristic), the large size of the dentary, the pattern of the cheek teeth, and the presence of a secondary palate.

The second line includes the family Scaloposauridae, fairly small reptiles (head measuring 8 to 9 cm in length) which lived in the Permotriassic and lower Triassic. Their mammalian characteristics are less marked than those of the Bauriidae. The secondary palate is only beginning to form from the two maxillae, which come into contact with each other. The postorbital bar is incomplete (a rather unimportant characteristic). The parietals are well developed. The pineal opening as well as the prefrontal have disappeared. The angular is small. The otic region is fairly similar to that of mammals.

Close to the Scaloposauridae is placed a family consisting solely of the genus *Ericiolacerta* (*Lystrosaurus* zone) from the lower Triassic, which is known from the works of Watson (1931). The characteristics of this small reptile (skull 4.5 cm long) show that its evolution was clearly directed toward specialization. The palatines contribute to the formation of a complete secondary palate; the pineal opening is missing; there is a single occipital condyle; the base of the skull is of the "bauriamorph" type; the presacral vertebral column is not differentiated into dorsal and lumbar regions.

Long and slender limbs seem to allow running. The foot is remarkable for its calcaneum, which is prolonged backward by a strong tuber calcanei, a characteristic structure of the mammalian foot. *Ericiolacerta* may well have been a digitigrade. The scapular girdle is reptilian, but shows a vestige of the cleithrum.

The latest bauriamorphs, however, bear a greater number of archaic features than contemporary cynodonts. For example, their lower jaw is of the gorgonopsian or therocephalian type, in which the mass of internal adductor muscles of the lower jaw is not differentiated into the masseter and temporalis mammalian muscles as in cynodonts (Barghusen, 1968).

ICTIDOSAURS, OTHER "PREMAMMALIAN" REPTILES

The Bauriamorphs gave rise to an interesting form: *Diarthrognathus broomi*, which is the only member of the infraorder Ictidosauria. An incomplete skull was found in the upper Triassic of South Africa.* The remarkable and important new feature of this theriodont is the double articulation of the jaw with the skull.

The articular bone, as in all reptiles, articulated with the quadrate in a troughlike depression; the quadrate articulated in turn with the squamosal, without a greater reduction of postdentary bones; a long process of the dentary, already visible in cynodonts, also articulated with the skull in a depression of the squamosal, the glenoid fossa. Thus, two articulations with the skull are present in the jaw, a reptilian one and a mammalian one. The coexistence of two characters, one (the new one) "meant" to replace the other (the archaic one), is a rare enough phenomenon to merit special attention. It is important to note that the mammalian articulation is entirely independent from the reptilian articulation. The entire evolutionary process has nothing to do with mutation.

A thorough study of the available parts of the skull of *Diarthrognathus* (still shallow, tapered anteriorly, broad at the back) has revealed a mixture of mammalian and reptilian characters.

The main mammalian features are the presence of a mesial wall in the

*The origin of ictidosaurs is still a matter of controversy. Hopson and Crompton (1969) believe they were derived from cynodonts, without defining their ancestry further.

orbit which is formed by the frontal and the palatine, the loss of the prefrontal and postorbital, the reduction of the transverse ramus of the pterygoid, the predominance of the dentary over all the other bones of the jaw, and the presence of a dentary–squamosal articulation and of two occipital condyles. Since the anterior part of the skull is missing, it is not known whether *Diarthrognathus* possessed a secondary palate.

THE EVOLUTION OF THE JAW MUSCULATURE

The musculature of the jaw, which was studied more specifically by Barghusen (1968), is of considerable interest, since it evolved in correlation with the jaw skeleton; the functional condition of the jaw depends directly on this musculature.

Theriodonts, except for the Cynodontia, possessed a single adductor muscle mass which had its main insertion (1) in part, on the sagittal crest and on the fascia covering the temporal fossa (but not on the zygomatic arch, except for a possible insertion on its posterior tip), and (2) in part, on the dorsal ridge and internal faces of the coronoid apophysis and of the supraangular. In these theriodonts, the apophysis was very close to the zygomatic arch so that the muscles did not pass underneath to insert into the lateral external face of the dentary.

This evolution has been slow and gradual. Thus, in Permian cynodonts, the zygomatic arch is much more curved than that of contemporary bauriamorphs; a wide space is present between the zygomatic arch and the coronoid apophysis which bears on its external upper half an excavation for the attachment of a muscle. It is the first indication in theriodont reptiles of the presence of a muscle reminiscent of the masseter, and inserting on both the zygomatic arch and the lateral external surface of the dentary.

In the best-known cynodont of the lower Triassic, *Thrinaxodon* (belonging to the Galesauridae), the masseter fossa has undergone considerable enlargement. It extends onto the lower part of the dentary; from this stage on, the masseter is as developed as in mammals. In correlation with this extension of the muscle, the dentary bone stretches backward much more than in the Procynosuchidae and develops a stronger coronoid apophysis as well as an increased angular bend.

In tritylodonts, the internal head of the masseter is entirely attached to the dentary, where its insertion forms a definite excavation. Thus, during the whole history of the therapsids, the ancestors of mammals, the development of the bony parts of the mandible (i.e., lower jaw) and of the motor elements—the muscles—has been closely coordinated. Such was also the case for innervation. The transformation of the reptilian mandible into the mammalian mandible could only occur thanks to a triple coordination simultaneously involving bones, muscles, and nerves. This is what we call evolution. It is not a mosaic of random variations affecting just anything at any time.

SUMMARY OF THE MAMMALIAN CHARACTERS OF THERAPSIDS AND OF THEIR ANCESTORS, THE PELYCOSAURS

In order to facilitate the reading of the material that follows a résumé is presented here of the main mammalian features of pelycosaurian and therapsid reptiles:

> *Order Pelycosauria* (from the upper Carboniferous to the lower Triassic): Regression of the supratemporal; mitration of the nares toward the anterior end of the snout; thinner, bifid stapes articulating with the quadrate in contact with the tympanum. They have been considered as the ancestors of the therapsids, which gradually replaced them.
>
> *Order Therapsida:* Gradual appearance of mammalian structures: secondary palate, rodlike stapes, formation of a powerful zygomatic arch (jugal and squamosal), etc.
>
>> Suborder Phtinosuchia (beginning of the upper Permian of Russia): Highly developed maxilla; long humerus; reduced rodlike stapes.
>>
>> Suborder Theriodontia: Dominance of the dentary; reduction of postdentary bones and quadrate; tendency toward heterodonty (incisors, canines, cheek teeth).
>>
>> Infraorder Gorgonopsia (upper Permian of Africa, Australia, and Russia): reduction in the number of cheek teeth; posterior elongation of short maxillae; fused vomers; straightened, slender limbs.
>>
>> Infraorder Cynodontia (from the late upper Permian to the middle and perhaps upper Triassic; southern Africa, Russia, China, South America): complete secondary palate formed by extensions of the maxillae and palatines: posterior elongation of short premaxillae; two occipital condyles, but small in size and very close to each other; no postfrontal; small pineal foramen; epipterygoid joining the wall of the neurocranium; an alisphenoid instead of a pleurosphenoid; ascending of ramus of the dentary

beginning; strong coronoid apophysis; beginning of the differentiation into dorsal and lumbar ribs.

Infraorder Tritylodontia (upper Triassic of South Africa, Europe, Asia, North America): Absence of prefrontal, postfrontal, postorbital; two or three rooted cheek teeth; two occipital condyles; internal auditory meatus; platicoelous vertebrae.

Infraorder Therocephalia (upper Permian of southern Africa, Russia): Phalangeal formula 2,3,3,3,3; missing postfrontal in *Euchambersia.*

Infraorder Bauriamorpha (from the upper Permian to the lower Triassic of southern Africa): Reduction or loss of the postorbital bar; secondary palate (in advanced species); occipital condyle tending to become double, single in some species (*Ericiolacerta*).

Infraorder Ictidosauria (Infraliassic of southern Africa): Double mandibular articulation; prefrontal and postorbital absent; reduction of the transverse ramus of the pterygoid; inversion of the articular surfaces of the posterior bones of the jaw; frontal and palatine contributing to the formation of the mesial orbital wall.

Paleontology shows that the succession of the characteristics of a group appears in chronological order. Thus in theriodonts are found characteristics of the vertebrate subphylum (excluding the more ancient ones which were acquired before the vertebrate condition was reached, those of the class (reptiles), of the order (therapsids), and finally those of the infraorder (terminal characteristics of the line). We will set up a separate category to include the erratic mammalian characteristics present in the various lines, families, etc., of theriodonts.

As in the case of ontogenesis, the most general characters are those which appear first (characteristics of the phylum: structural plan), followed by those with signs of specialization, and finally the characteristics of the family, of the genus, and of the species. Figure 4 shows the theoretical relationships among the main branches of theriodonts.

CR: General characters from the ancient base of the vertebrates and from reptiles (controlling, notably, the fundamental cellular mechanisms); they are constant.

CM: Mammalian characters which appear sporadically. They can vary. The greatest number of mammalian characters are found in the infraorders Tritylodontia and Ictidosauria.

CL: Lineage characters, constant.

The anomodonts, which are presumably derived from the Phtinosuchia (from the middle Permian to the lower Triassic), form a complex unit of vegetarian reptiles which have not evolved in the mammalian direction. They need not be considered here.

Kühne (1956) listed ten mammalian characters found in the therapsids of the middle Permian and lower Triassic (excluding anomodonts): (1) olecranon; (2) tuber calcis; (3) epiterygoid plate; (4) secondary palate; (5) double occipital condyle; (6) obturation of the pineal opening; (7) loss of the postorbital bar; (8) obturation of the ectocondylar foramen; (9) free acromion; (10) phalangeal formula 2,3,3,3,3. The extension of the dentary and the reduction of the postdentary bones, quadrate, and squamosal should, in our opinion, be added to this list.

But none of the theriodonts shows all these characters; *Diademodon*, the best known cynodont, only shows five of them, and *Cynognathus*, four. The tritylodonts bear such additional characters as cheek teeth with two or three roots, an internal auditory meatus, the loss of the prefrontal and the postorbital, platycoelous vertebrae, ribs articulating with the vertebrae as in eutherian mammals, an odontoid apophysis (axis), cervical ribs fused with vertebrae, a preacetabular ilium, and a cotyloid notch.

The data supplied by paleontology and comparative anatomy enable us to trace the gradual evolution of the mammalian cranium. It began toward the end of the Carboniferous with the appearance of pelycosaurs and came to an end in the Rhaetic (lower Jurassic), lasting for approximately 100 million years. It involved several phenomena which are closely correlated, namely:

1. The cranial bones decrease in number; they either disappear or become fused. The septomaxillae, prefrontal, postfrontal, postorbital, supratemporal, and quadratojugal have disappeared. The parasphenoid vanishes or becomes vestigial in some species; the postparietals and the tabular become fused with the occipital; the epipterygoid becomes fused with the skull wall.

2. The relative sizes of the various bones change, for instance, in mammals, the upper maxilla is greatly enlarged and reduces the size of the lachrymal (the proportions vary from one subclass to the next).

3. The arrangement of the bones is modified by size changes, losses, and fusions.

4. In mammals, the parietals and frontals tend to form the major part of the skull roof; they lie between the temporal jaw muscle and the brain, thus completely shutting off the neurocranium. The occipital condyle becomes double. The buccal cavity is transversely partitioned by a secondary palate*; the teeth are differentiated into three categories, from

---

*The secondary palate (or secondary roof of the mouth) is formed by four apophyses arranged in horizontal laminas sent out toward the internal face by the upper maxillary

front to back, the incisors, canines, and postcanines. To the primitive simple conical or thornlike type are added extra points or cusps in longitudinal rows. The teeth can thus chew instead of lacerate. The upper teeth are located on the premaxillae and maxillae, the lower teeth on the dentary bones.

5. The evolution of the lower jaw has followed the *same* trend in *all* phyla. It is characterized by a reduction of the postdentary bones and a considerable development of the dentary, and by changes in the positions of muscle attachments and nerves; these changes occurred simultaneously and were closely correlated. The vertebral column is greatly modified. It becomes differentiated into two regions: a dorsal region bearing elongated ribs and a lumbar region bearing curved ribs immovably attached to the vertebrae.

A COMPARISON OF THE RESPECTIVE CHARACTERS OF MAMMALIAN REPTILES AND EARLY MAMMALS: GRADUAL TRANSITION FROM ONE CLASS TO THE OTHER

With the appearance of *Diarthrognathus* and *Oligokyphus*, the mammalian structure begins developing but is far from being perfect. This can be better appreciated by considering the difference between the skull of the most advanced theriodonts and that of mammals, which is remarkable for the great expansion of the braincase, for the lower jaw formed by a single bone, the dentary, for the absorption of some postdentary bones into the chain of ossicles of the middle ear, for the reduction of the postoccipital process, and for a few other minor differences, namely:

1. The nasal pharyngeal structures pushed far backwards.
2. Reduction of the pterygoids and their separation at the midline.
3. The reduction of the ectopterygoids, their extension backward, and their inclusion into the wall of the nasopharyngeal duct.

---

and palatine bones. This anatomical structure does not characterize mammals exclusively, since it is also present in crocodiles, in which it consists of palatine apophyses sent out by the premaxillary, maxillary, palatine, and pterygoid bones. It is perforated anteriorly by the premaxillary foramen and laterally by the palatine foramina. The vomers contribute to the formation of the palate of crocodiles of the genera *Melanosuchus* and *Tomistoma*.

This secondary palate, therefore, appears in a group which, in other respects, shows no affinity with mammals. The separation between the oral cavity and the nasal cavities would also have been a useful adaptation in the other groups of reptiles, particularly in aquatic forms. The formation of the secondary palate has been related to food mastication and to heterodonty (see Vanderbroek, 1969, p. 256). Such a relationship cannot be said to exist in crocodiles. One can note that several anomodont genera (notably dicynodonts) possess a highly developed palate in the anterior region of the roof of the mouth; yet, they are not considered to be mammalian ancestors.

4. The shortening of the brain case floor in the temporal orbital region, the palatines and basipterygoid processes being brought closer.

5. The loss of the parasphenoid; preservation of the presphenoids and basisphenoids in the ventral view.

6. The anterior root of the zygomatic arch frequently pushed forward.

7. The great reduction of the squamosal.

This enumeration could well be continued, but we will stop here. It demonstrates effectively the break which still separates the most mammalian reptiles from true mammals.

It used to be easy to distinguish reptiles from mammals. An examination of the articulation of the jaw was sufficient: in reptiles, the articulation is between the articular and the quadrate; in mammals, between the dentary and squamosal. But the discovery of reptiles bearing both kinds of jaw suspension makes the point of distinction harder to define. In fact, classifying certain fossils in one class or the other is akin to scientific aberration, because they are both reptilian and mammalian. Their systematic position in the classification can thus be arbitrary only.

In effect, the paleontologist knows really only one aspect of the being which he is trying to reconstruct. He has no knowledge of the soft parts and their functions. Not knowing whether these beings were piliferous and "mammalian" (in the etymological meaning of the word), how could he express an opinion with any degree of certainty?

Among the characteristics available for distinguishing between reptiles and mammals, one of them, discovered by Hopson (1971), seems to us to be most significant. Mammals only have two kinds of successive visible dentition (they are fundamentally diphyodont), whereas all the theriodont reptiles so far studied have a greater number of successive dentitions (they are all polyphyodont). Searching for facts bearing upon selection, Hopson suggests that mammalian diphyodontia is strictly correlated with the dependence of the young on their mother for care and feeding, whereas in reptiles (including cynodonts) the young, when hatched, are independent and self-sufficient. We find little relationship between the number of dentitions and the feeding habits of the young. Hopson tends to forget that mammals can easily do without lacteal dentition: toothed whales are monophyodont; the functional dentition of sirenians is lacteal; the molars of the second dentition of the dugong are resorbed before they ever function.

Moreover, from a Darwinian utilitarian standpoint, one can regret that mammals only have one replacement dentition: For instance, it is an inconvenience to man, and the short life span of some rodents (Microtinae, for example) is due to the rapid wearing down of their teeth, which do not grow back. But what can be done about it?

THE CHARACTERISTICS OF CREATIVE EVOLUTION IN THERAPSID REPTILES

By considering the characteristics acquired by theriodonts in over 60 million years, their respective arrangements, their chronological order, we can draw the following conclusions:

1. The evolution of theriodonts was oriented and progressed in two directions: in a general mammalian direction; and in a direction specific to each line.

2. Mammalian characters are not the same in all the theriodont lines, nor are they expressed in equal or identical manner in all. Each line imposes, to a greater or lesser extent, its distinctive style.

3. The variations undergone by the theriodont reptiles accumulate with the passing of time (this idea is expressed by the term "tend to").

4. The variations are complementary and when they appear induce coordinated variations, such as the development of the dentary with concomitant transformation of the jaw muscles.

5. The variations which are observed in the theriodont lines are orderly. They never occur in a chaotic manner.

6. Transformations in lines and sublines occur very gradually, as a continuous process, without radical changes. Everything is weighed out, equilibrated. Incipient structures strengthen and gradually become well formed and functional organs.

7. At no time did innovations affect a fetal character; at no time were fossil forms arrested in their development. This evolution has nothing to do with neoteny, with fetalization, as understood by Bolk.

Most of these facts have been partially observed by paleontologists (Romer, Hopson, Crompton, and others) who mention mainly parallel evolution and allude to neo-Darwinian theory; their speculations on evolutionary phenomena hardly go any further.

The family tree of theriodonts is bushlike, and all its unequal branches are directed toward mammalian structure, following varying ways. According to Darwinian doctrine, this persevering orientation of evolution would depend on preexisting information contained in the strands of chromosomal DNA and stemming from a series of random events.

Before outlining the relevant information produced by paleontology, we wish to restate briefly (as it is so evident) that Darwinian factors, i.e., random mutations and subsequent selection, played no part in the evolution of theriodont reptiles. Yet Olson (1944), Simpson (1960), and Mayr (1963) held that these factors do explain the evolution of reptiles. They all postulate that the Darwinian doctrine is true. They only take into account the facts that fit in with the theory, strictly excluding those which indicate its weaknesses.

In fact, we quite objectively note that the variations of theriodonts and early mammals did not occur at random; they keep accumulating and adjusting in time and lack any pathological character. No recourse to selection is required to explain any feature which characterizes "mammalization." On the contrary, the diversity of subtypes (evolution is branching), the large distances between populations, and the variety of the climates to which they are submitted do not support the theory of selection.

The account given by G. Simpson of the history of the Equidae exemplifies the influence of this doctrine. "Unconditional" Darwinians deny the existence of oriented evolution and mistake it for orthogenesis, but nevertheless constantly use it unintentionally. Simpson concurs with the classic opinion according to which all Equidae derived from the genera *Hyracotherium* (Eocene of Europe) and *Eohippus* (Eocene of North America), which, to him, are synonyms, and which gave rise to two lines: true Equidae in North America (which crossed over to Europe in the Pliocene-Pleistocene) and the Palaeotheriidae in Europe. We will deal here only with the Equidae which have radiated into small evolutionary branches, such as *Anchitherium, Hypohippus, Stylohipparion*, and into true Equidae which ends with the genus *Equus*. According to Simpson, the diversity of lines proves the nonexistence of evolutionary orientation; yet, the process that he describes is precisely branching or splitting evolution, which is the *leitmotiv* of the equine form in all the sublines; it is more or less successful, but always present. Within each line, with the passing of time, the attainment of the equine idiomorphon is progressing. The picture of evolution that Simpson (1951) presents in his book "Horses" suits us. He depicts very well the evolutionary trends and writes about "orthogenesis."

The *only* abruptly appearing characteristic which Simpson reports is the filling with cement (bony tissue) of the grooves between the enamel ridges on the grinding surface of the cheek teeth in *Merychippus* (middle Pliocene of North America). In fact, this character had appeared earlier, in Europe during the Oligocene, in *Plagiolophus javali* (belonging to European Palaeotheriidae) and is one of the equine characteristics; in *Plagiolophus* it is associated with hypsodonty and with a marked tendency toward monodactyly: the middle toe (III) predominates over the two remaining side toes.*

Such an array of characters occurring on both sides of the Atlantic (earlier in Europe than in America) rules out the presumed incidence of chance. The populations of equids of Europe and America provide a "natural" experiment from which instructive conclusions can be drawn. It cannot be cited in support of Darwinian theory.

In sum, in all hippomorph lines, the following tendencies can be recorded in the successive genera: The general size increases; the facial mass increasingly gains importance over the neurocranium; there is an increase in the height of the teeth (cement develops) and in the complexity of the enamel folds; the limbs increase in length; the ulna and fibula become vestigial and fused with the radius and tibia, respectively; the predominance of the third toe is a constant fact: it began appearing in families which, although tridactylous, were functionally monodactylous and became fully expressed in the true monodactyly of the Equidae. The equine idiomorphon has a great many other peculiarities, such as a characteristic brain structure, skull suspension, general body line, etc., which appear in varying degrees in the sublines, and are thus "personalized."

In his theoretical discussion, Simpson does not linger over the structure of the hoof; yet it is the result of a very innovative and precise evolution. Such a hoof, which is fitted to the limb like a die protecting the third phalanx, can without rubber or springs buffer impacts which sometimes exceed one ton. It could not have formed by mere chance: a close examination of the structure of the hoof reveals that it is a storehouse of coaptations and of organic novelties. The horny wall, by its vertical keratophyl laminae, is fused with the podophyl laminae of the keratogenous layer. The respective lengths of the bones, their mode of articulation, the curves and shapes of the articular surfaces, the structure of bones (orientation, arrangement of the bony layers), the presence of ligaments, tendons sliding with sheaths, buffer cushions, navicular bone, synovial membranes with their serous lubricating liquid, all imply a continuity in the construction which random events, necessarily chaotic and incomplete, could not have produced and maintained. This description does not go into the detail of the ultrastructure where the adaptations are even more remarkable; they provide solutions to the

---

*The trend leading to the equine idiomorphon was continued in palaeotheres, which disappeared too early to acquire complete or even functional monodactyly.

problems of mechanics involved in rapid locomotion on monodactyl limbs.

THE MAMMALS: A HOMOGENEOUS OR A HETEROGENEOUS CLASS?

The latest data collected by paleontologists suggest that mammals are derived from cynodonts, and that the latter are related to the Ictidosauria. But, as noted above, cynodonts have radiated into various branches, several of which probably attained the mammalian stage. The fact that ictidosaurs and tritylodonts are almost "mammalized" supports this view.

The first mammals appeared at the end of the Triassic and were still bearing the signs of their reptilian ancestry. They have been found in Europe, Asia and Africa. They belong to three families,* the best known being the Morganucodontidae (*Morganucodon* of the Rhaetic of Glamorgan, Wales).

The lower jaw, consisting of the dentary only, bears an articular condyle which articulates with the squamosal. On the internal side it shows a longitudinal crest running along a deep groove. In *Oligokyphus,* this groove contained Meckel's cartilage, which was ossified on its proximal end and formed the articular. Hence, one must conclude that *Morganucodon* and allied genera possessed an articular which had not yet become the hammer (malleus) of the middle ear, but no longer formed part of the jaw articulation. This pattern seems likely, since in the groove of another specimen were found bony fragments which belonged either to the prearticular or to the articular.

The embryogeny of present mammals, suggestive of former states, supports this interpretation. Meckel's cartilage (distal part of the mandibular arch of the embryo) becomes in its distal portion the lower jaw (the reptilian dentary) and in its proximal portion, the head of the hammer (the reptilian articular) and the handle of the hammer (the reptilian prearticular). The quadrate in reptiles is derived from the palatoquadrate cartilage which overlies Meckel's cartilage; it becomes the anvil (incus) in mammals. The proximal end of the hyoid arch (Reichert's cartilage in the embryo) becomes the stirrup (stapes). Anvil, hammer (malleus), and stirrup from the chain of obsicles of the middle ear. The position of the jaw skull articulation in mammals corresponds to that of the dentary–articular articulation in the reptilian ancestors.

*The Haramyidae, known only from isolated teeth, cannot be precisely classified.

Morganucodon had 48 to 54 teeth composed of incisors, premolars, and molars (5/5 I; 1/1 C; 3/2 Pm; 5/5 M) (I).* They possessed mammalian and reptilian characters (Mills, 1971): the almost perfect dentary occlusion and the sectorial structure of the jugal teeth are mammalian; the number of teeth, the presence of an incisiform tooth on the maxilla anterior to the canine, the replacement of cheek teeth from the posterior end, and the early appearance of posterior molars are cynodont characters.

In sum, the teeth of *Morganucodon* are very similar to those of triconodont mammals, and Vanderbroek (1964) and Hopson and Crompton (1969) have considered this genus to be the most primitive triconodont. In both, the cheek teeth bear sharp triangular cusps with extra points; they look like the cynodont teeth, but foreshadow the tribosphenic teeth of therians.

The resemblances between the skeletons of the Morganucodontidae and monotremes have been greatly emphasized: reduction of the alisphenoid, of the petrosal bone widely contributing to the formation of the lateral wall of the cerebral skull; the pectoral girdle bearing a coracoid, a procoracoid, and an unpaired bone, the interclavicle.

However, most paleontologists, considering the skull characters (alisphenoid of the cynodont type; well-developed lachrymal and jugal), believe that *Morganucodon* of the Rhaetic (Glamorgan, Wales), *Docodon* (Morrison formations, Wyoming), and *Peraiocynodon* (Purbeckian of England) merit placement in a special order, the Docodonta. Since we have only incomplete skeletons and teeth, it is very difficult to define with any degree of accuracy the phyletic links between the various groups of Triassic, Jurassic, and Cretaceous mammals.

Triassic and Jurassic mammals preserve several reptilian characters. For example, the Dryolestidae, a family of early eupantotheres (Kimmeridgian of Portugal) possess in their lower jaw a splenial and Meckel's cartilage which persist in the adult form; these reptilian elements are lost in the last Dryolestidae (lower Cretaceous of Spain) (Krebs, 1971). We may add that these therian mammals should probably be excluded from the ancestry of eutherians (higher mammals).

The fact that the lateral wall of the braincase does not have the same composition in all mammals (Kermack and Kielan-Jaworowska, 1971) is most important. In fossil and living therians, it is formed by the alisphenoid and the squamosal. In the nontherians (docodonts, triconodonts, multituberculates, monotremes), it is formed by the anterior blade of the petrosal.

*In the earliest therians (eupantotherians), the number of teeth is still very high: the lower jaw of *Amphitherium* (Bathonian of England) bears 32 teeth, and the lower jaw of *Dryolestes* (*Loalestes*) (Morrisson formation, Wyoming) bears 34.

It is possible that this structural difference may have a diphyletic origin; the therian and nontherian would be derived from two separate groups of Jurassic theriodonts. There is no reliable proof of this, however.

Resemblances have been reported between monotremes and multituberculates: In both, the alisphenoid is reduced, the jugal is missing, the lacrimal is very small or absent, the ectopterygoid of the multituberculates is similar to the pterygoid of echidnas; but does this imply common derivation? The specialization of the second form is so marked that this is questionable. Their dentition, which is roughly that of rodents, raises further doubt. The belief that multituberculates form a line independent from other mammals is based on anatomical arguments concerning the differences between their skeleton and that of monotremes.

In monotremes, the main reptilian characters can be found in the pectoral girdle (procoracoid, coracoid, interclavicle) lying anterior to the thoracic cage,* in the structure of the lateral wall of the braincase, in the laying of large telolecithal eggs, and in the existence of an eggtooth (*ruptor ovi*) placed, as in lizards and snakes, on the premaxillae. The primordial (i.e., mucocartilaginous) skull of monotremes possesses the original wall (pila antotica) found in the reptilian skull.

Such an array of facts suggests that the small group Monotremata (arrested very early both in the course of its evolution and of its expansion by peculiarities in structure and reproduction) separated very early from the other mammals; the monotremes are presumably descended from some Mesozoic mammals that evolved from a line of cynodonts different than that from which the Metatheria and Eutheria evolved. Although several paleontologists believe that multituberculates were the ancestors of monotremes, no reliable conclusion can be reached since we have no ancient fossil evidence of the latter.

The case of marsupials is just as ambiguous, although their anatomy and embryogeny show numerous peculiarities. Zoologists and paleontologists generally agree to give them the same ancestry as eutherians, that is a pantothere with tribosphenic teeth and prismatic enamel. Our ignorance of the soft parts makes these conclusions somewhat questionable, but it is likely that both groups diverged very early from the common panthothere block, probably in the middle Cretaceous.†

---

*Howell (1937, p. 191), an authority in such matters, wrote: "The shoulder of monotremes must be considered as the expression of a highly specialized reptilian type."

†Simpson (1928, 1959), Olson (1944), and a few other paleontologists supported the idea of the polyphyletic origin of therians. In the last few years, a reversal of opinions has taken place: Hopson and Crompton (1969) would favor monophyly, whereas Kermack (1967) prefers diphyly. Romer (1970) remains cautious and does not clearly adopt either conclu-

The heteregeneous character of the class of living mammals is obvious. One can reasonably think that the first mammals were not all descended from the same line of theriodonts and that monotremes have an ancestry distinct from that of therians, but their monophyletic or polyphyletic origin has little bearing on our discussion. What is essential here is the diversity of orders and suborders, which favors the attainment of the mammalian idiomorphon; this organization does not involve rectigradation or orthogenesis and occurs gradually along numerous paths from a common archaic stock. Evolution is oriented, there can be no doubt about it, but it follows varied routes: All the subgroups lead to the idiomorphon. The history of mammals is proof of this.

Mammalian evolution, as soon as the skull acquired its ultimate structure, was characterized by a considerable increase in the volume and the complexity of the skull. One must insist upon the fact that in theriodonts, "mammalization" transformed mainly the skeleton, the musculature, the dentition, but hardly touched the brain. It was fully achieved only in true mammals, in which the rise of "cerebralization" has been the major evolutionary theme. The neurocranium, even in orders with a powerful facial mass, never undergoes reduction; the case of perissodactyls is very instructive in this respect.

CONCLUSION

Judging from the fossilized remains, creative evolution possesses its specific characters distinct from those of mutagenesis (see Chapter IX).

Mutations do not explain how coordinated variations play upon several organs at a time; some lethal or sublethal mutations of a gene produce multiple effects. Thus, rhinoceros mice are the result of the mutation of a single recessive rh gene—shortly after birth they lose their hair; at the age of one month, when hair should normally grow back, their skin wrinkles and loses its elasticity. This change is seemingly due to the formation of numerous intradermal cysts from isolated fragments of the hair bulbs or follicles. The animal's nails grow and curl up; the entire body is covered by a waxlike secretion. The average life-span of the mutants is half that of normal mice. The rh gene, in homozygous individuals, prevents the metabolism of lipids, induces a deficiency of triglycerids, a considerable accumulation of waxy products and a higher

---

sion. Simpson (1971), without mentioning his previous opinion, dwells on the definition of mammals and on the period of appearance of the first or first few representatives.

percentage of an ester, lathosterol. It therefore simultaneously acts upon several organs.

Another example of this can be given. A mutation substituting one or more amino acids for one or more others in the globin of human hemoglobin may, depending on its location, have serious effects on the structure (at various levels) and properties of this pigment; such as the case of abnormal hemoglobin S in the anemia of cresentic red blood cells (drepanocytosis or sickle cell anemia). The molecules of this hemoglobin line up in filaments and form stiff rods, grouped into bunches which distort the blood cell. The red blood cells containing hemoglobin S are practically destroyed, and the resulting anemia causes serious troubles—such as the increase of the bone marrow which, in order to speed up the genesis of red blood cells, modifies the skull bones; the hypertrophy of the heart; and the accumulation of crescentic blood cells in the spleen, which increases in size and becomes fibrous.

But these pathological cases have no relationship whatsoever with the slow and coherent attainment of a new form, of a new function. We do not question the existence of the multiple effects (i.e., pleiotropy) a single gene can produce but we do not think any geneticist would maintain that the transformations of the jaw, of its muscles and nerves, of the ossicles of the middle ear, etc., could be induced by a single gene.

The pleiotropic mutation, from what we now know about it, causes the loss of one enzyme, which results in a number of consequences (such as the eye pigment of *Drosophila*, the synthetic ability of the fungus *Neurospora*). This mutation does not construct anything; it alters or destroys preexisting elements.

Incidentally, I believe that the application of genetics to the study of fossils is a difficult and perilous task. Since no breedings are available, what criterion will determine whether a given change, recorded on a bone or a shell, is indeed a mutation? Paleontological genetics, even in such a favorable case as that of the Paludinae of Slavonia, cannot avoid being subjective and hypothetical.

Mayr refers to natural selection and maintains that selection pressure was strong in the various forms of theriodonts. How does he know? His conclusion does not rely on any demographic data (we have no knowledge of population densities, ecological factors, etc.), nor on ecological or climatic data. We know little about the environments in which theriodonts lived. How could natural selection give rise to the single mammalian structure while it acted upon populations living in very different environments (Asia, South Africa, South America)? Environmental conditions change from one continent to the other, and the climates were already characterized both in the Triasic and in the Jurassic. How could natural selection, in the midst of such diversity, manage to favor the same forms everywhere, without being inconsistent with the neo-Darwinian principle that "each environment has its own privileged genotype which, by chance, is better preadapted to it"? We wonder.

III

Evolution—A Discontinuous Historical Phenomenon

To understand a forest, its life and biological cycle, two methods are open to us: one can either undertake a detailed analysis, an inventory of the different species tree by tree, a study of the texture and chemical composition of the soil, a study of the climate and the biotopes of the area, or one can consider the forest in its totality, as a superorganism with its own particular life and ecological characteristics. Similarly, one can follow evolution line by line and even occasionally genus by genus, or compare it to a huge fresco whose deepest meaning can only be fully understood if looked at from a distance. In fact, biologists have no choice: they must have recourse to both methods because they are complementary. I intend to provide evidence for this in what follows.

EVOLUTIONARY DISCONTINUITY

It is generally presumed that evolution is a continuous phenomenon which began with the synthesis of prebiotic substances and will end with the death of the last organic being on an aged planet unfit to sustain life any longer. A number of biologists accept this theory as a general law which they call the *law of evolutionary continuity*. As things happen in nature, it is both true and untrue that evolution is a continuous phenomenon. Let us explain this further.

Seen as a whole, the living world appears unstable and changing; the longer the period considered, the larger the modifications appear. If considered in detail, one discovers that the various lines of descent do not all change at the same rate.

Evolution is a phenomenon which occurs at varying rates and even stops in many lines of descent. In nature at present, evolution occurs in moderation; a great many organic or physiological changes are wrongly attributed to evolution. Let us examine two extreme cases: the multiplication of bacteria and that of hominids.

A human generation lives an average of 25 years; a generation of bacteria, half an hour (20 minutes at 37°C). In other words, multiplication in the second case is at least 400,000 times faster than in the first.*

Let us consider the period of 3,500,000 years between the early Australopithecines and modern man. During this interval, which is long enough to reveal notable evolution, mutations appeared in bacteria and also in hominids.

If one refers to data produced by bacteriologists (Stanier *et al.*, 1966), populations of cultivated bacteria produce one mutant per 10^9 self-dividing individuals. This rate seems low compared to that of cellular beings; it is presumed that it is higher in natural populations of bacteria.

Whatever the actual rate may be, the immensity of bacteriological populations is such that the number of mutants is likewise enormous. Over a period of 3,500,000 years, this number exceeds billions of billions. During the same lapse of time, the smaller populations of hominids gave rise to a few million mutants at the most.

Bacteria, reproducing at a rate of 17,520 generations per year, existing in the most varied environments (e.g., in the bodies of animals and plants, in soil and water), and all producing numerous mutations, have in no way changed their structural plan which has remained the same since the time when their Precambrian ancestors were metabolizing ferric salts in the lagoons. Hominids, during these 3,500,000 years, continued their rapid evolution, perfected their brain structure, became quite bipedal and straightened to an erect stature. Such a comparison shows us that mutations do not necessarily lead to an evolution, no matter how great their numbers are; this comes as a severe blow to ultra-Darwinian ideas.

Schizophytes† and others of the same type have existed since extremely remote times, as far back as the Precambrian period. The main

---

*We are not referring to closed bacterial populations with no new input, but to natural populations in which reproduction is not slowed down or inhibited by the depletion of food supply and the presence of excreta in the environment. Such unfavorable conditions often result in the death of the populations.

†Schizophytes, which lack a nucleus and reproduce by binary fission (without mitosis), include the true bacteria or Eubacteriales (*Bacillus, Vibrio, Spirillium, Pseudomonas, Actinomyces*, etc.), and also the Myxobacteria, spirochetes, and blue-green algae or Cyanophyceae.

historical fact to be remembered is that, in spite of their infinite number, they have only *once* given rise to a cellular organism, following a process about which we know *nothing*.

One must keep in mind that a cell assumes the same pattern in all organisms and contains a nucleus with an enclosing membrane, perforated by a great number of pores in which are enclosed DNA macromolecules incorporated into organelles, the chromosomes, which are highly organized and whose structure comprises various proteins, histones being the only ones on which extensive research has been conducted.

In nature, at present, schizophytes and among them especially bacteria—because of their great number of mutations—vary extensively and in the greatest disorder; but they "go around in circles" revolving around their specific form, and in the end change very little, that is to say, do not evolve. Bacteria have remained bacteria for the past two or three billion years.

Four events have had an immense effect on the evolutionary path: these are the synthesis of chlorophyll, the change from schizophyte to cell, from a single cell to a multicellular organism, and from diploblastic metazoans to triploblastic metazoans. Sexual reproduction, by combining in a single being the characteristics of two, plus the characteristics of the entire population through the interplay of the consecutive generations, has played a fundamental role in the history of the two kingdoms.

The genesis of the fundamental structural plans which characterize the subphyla and classes—themselves the main branches of the genealogical tree of the animal kingdom—has been the greatest achievement of evolution. There have been few creations: fewer than twenty phyla and eighty classes for the animal kingdom (less than half that many for the plant kingdom). They are all very ancient. The last major group to date, the vertebrates, made its appearance with the Agnatha (ostracoderms, Cyclostomata) during the Ordovician, and with jawed fishes during the Devonian, some 450 millions years ago.*

Since the Jurassic (Rhaetic, 200 million years ago), when the first mammals and the precursors of birds (Portlandian, 135 million years ago), no new classes have appeared.

The interruption of the genesis of the fundamental plans and the fact that they are so few in number are two facts that have been underesti-

---

*If the Conodontophoridae were proved to be chordates, the date of appearance of the first vertebrates, or at least their ancestors, would have to be pushed back to the Cambrian. *Emmonsaspis*, which was found in the Parker slate (Cambrian of the Appalachians), would thus be an archaic cephalochordate. This seems unlikely.

mated in the attempt to understand evolution. Their meaning is as follows:

Evolution does not occur in just any fashion, since the number of its possibilities is limited.

The organisms in which the new plans were embodied have partly lost their evolutionary capabilities; these will not recur in their descendants.

The creative powers of evolution have gradually decreased with the aging of the flora and fauna, and since the Eocene (60 to 33 million years ago), the formation of orders has been interrupted, eutherian mammals and birds being the last to appear.

Arthropods were represented as early as the Cambrian by the trilobites, merostomes, and crustaceans. The first group became extinct in the Permian, after a long period of decline which began in the Devonian; the second group became extinct during the same period, except for a few horseshoe crabs (*Limulus* and *Carcinoscorpius*); as for the Chelicerata (other than the Merostomata) and the Crustacea, they have continued to evolve and to give rise to subclasses and orders.

Insects, which can be traced to the Devonian, have constantly remained numerous and varied. Like the Crustacea, some of their orders and superfamilies have indeed become extinct; however, their antiquity notwithstanding, they have always remained unchanged during the course of their history; they retain as many types as in the past.* Despite such "vigor," the history of the insects—the starting point of which is unknown—did not occur in regular sequence. Subclasses succeed each other in a quasilinear fashion: some reached a peak, then declined; some remained as flourishing as when they first appeared (such as the Collembola, Orthoptera, Odonata, and all the Holometabola). The Paleopterat were dominant from the Westphalian to the end of the Permian; they were then supplanted by the Odonata and Ephemeroptera, with numerous archaic features. The Polyneoptera (cockroaches, termites, Plecoptera, Orthoptera, Dermaptera), which were contemporaneous with them, have been flourishing and producing numerous species since the end of the Palezoic and still do not show any sign of exhaustion.

The Oligoneoptera (Coleoptera, Neuroptera, Mecoptera, Diptera,

---

*During the course of their long history, the class Insecta has lost three superorders: The Paleodictyoptera (comprising seven different orders), the Megasecoptera and Protoorthoptera, which lived during the Permo-Carboniferous period. All other superorders have survived and still thrive today.

tThe classification adopted here was set up by Martynov (1938). It is quite rightly based on data from paleontology as well as from comparative anatomy (see Jeannel, 1949a,b).

Trichoptera, Lepidoptera, Hymenoptera) and the Paraneoptera (Psocoptera, Thysanoptera, Hemiptera) are the orders which comprise the greatest number of species in the whole of the animal kingdom. Some of them, such as the Ledidoptera and, to an even greater extent, the Hymenoptera, have produced a spectacular "fireworks" of new types when, in the Cretaceous, flowering plants widely developed. From the onset, they occupied the new environment and radiated into a considerable number of species. Close relationships have been established between the flowers of numerous phanerogams and the nectar- or pollen-eating insects, to the extent that a few plants are exclusively pollinated by visiting insects which carry pollen from one to the other; this happens in orchids, which show numerous adaptations to this morphological, physiological, and ethological mode of reproduction.

The class Insecta includes some venerable relics: among the odonates, *Hemiphlebia mirabilis*, which lives in Australia, is very closely related to the Protozygoptera of the Permian. Modern cockroaches are little different from the cockroaches of the Paleozoic Era, when, due to the expansion of forest areas, they found a huge biotope which suited them. As early as the upper Westphalian, they laid eggs enclosed in an ootheca, which is precisely what the present species do. Since the Permian, ephemerids have changed very little; for instance, the larva of *Phthartus rossicus* of the upper Permian of Orenburg (Russia) are very similar to those of mayflies now found in our streams.

The Mecoptera have preserved very close relationships with their Permian direct ancestors such as *Taeniochorista* and *Chorista* of Australia, *Nannochorista* of Chile, Argentina, Tasmania, and Australia, and *Choristella* of New Zealand, which seem to be of unquestionably Gondwanian origin.

One can conclude from the above evidence that the class Insecta has never experienced any decline and is still flourishing, but that its creative stage is over. From the last 30 to 70 million years, it has not gained any new orders, and since the Oligocene (Amber of the Baltic) its evolution has not gone beyond the generic level. A great number of small variations (i.e., mutations) can be observed, for example, in *Drosophila*, *Sciara*, *Habrobracon*, and the Doryphora—they delight geneticists and are their bread and butter!—but these mutations do not alter the specific structural pattern. Most of the insects caught in the Oligocene resins belong to genera still existing; for example, three fossil species of the termite *Reticulitermes* (widespread in the palearctic zone at present) are known to us. Certain identifications may lack accuracy and genera may still be imprecisely defined, but it does not matter a great deal: the resemblances between the Oligocene insects and surviving species are so great as to clearly indicate the weakening of evolution.*

Mollusks, which, prior to the end of the Mesozoic, lost ammonites and belemmites; echinoderms, which suffered heavy losses (Cystoidea, Blastoidea) and possess a relict class (Crinoidea); and sharks and actinopterygian fishes are all thriving still, and relict forms coexist with recent forms. In all of them, equally or to a greater extent than in insects, evolution has become hardly noticeable. Excluding a number of gastropod groups (*Viviparus,* the Helicidae, *Partula, Achatinella*) and bivalve molluscs (unios, anodonts) in which the speciation related to neoendemism continues, they are the very picture of evolutionary stagnation.

The last major evolutionary "explosion" occurred after the disappearance of the dinosaurs, at the junction between the Mesozoic and Cenozoic eras, when the various orders of eutherian mammals and birds became differentiated. During that same period the marsupials (metatherians) of Australia underwent a similar evolution, occupying most of the ecological niches of that continent, and gave rise to forms looking like those of the eutherians prevalent on the rest of the globe: carnivorous forms (imitating the Canidae and Ursidae), climbers, flyers (parachuters), burrowers (of the mole type), and arboreal forms, but with a lack of forms imitating the primates.

The notoungulates, relict in South America from the Eocene to the Quaternary, have imitated ruminant artiodactyls, equoid perissodactyls, and rodents. They have all vanished without descendants.

Some very instructive data can be obtained from the known stages of the evolution of primates. In all primates, excluding man for the moment, evolution has been suspended for a long time. The tarsiers (genus *Tarsius*) are very closely related to fossils of the Eocene; the lemurs, the history of which is still partly unknown, exhibit forms specialized to unequal degrees; and the arboreal Lorisidae (continental Africa and East Asia) with hands and feet forming claws much like those of chameleons, day-blind, with a slow, silent, cautious locomotion, are undoubtedly those engaged to the greatest extent in specialization; however, all lemurs appear to be morphologically stabilized relics rather than evolving forms.

The Cercopithecidae are only known since the Miocene; they were then already engaged in the course which led them to the cynomorph stage fully attained at the junction between the Miocene and Pliocene.

---

*Insects of the Cretaceous age, judging from the fossils found in the Amber of Manitoba, are highly reminiscent of the entomological fauna of the Tertiary Era and of the Modern fauna.

Mesopithecus (Pontic or Greece) was an ape very similar to the sem-nopithecines (*Presbytis*) which now live in India. The evolution of the Cercopithecidae was thus effected over 10 to 12 million years ago. The ancestors of the pongid Anthropormorpha are the Dryopithecines (*sensu latissimo*) of the Miocene, and *Proconsul*, which dates back to the Pliocene (9 million years ago). The latter is so closely related to the two African pongids, the chimpanzee and the gorilla, that one might say that these two present-day apes existed already in the form of *Proconsul* in the Pliocene. We can thus affirm that the three major lines, the tarsiers, lemurs, and simians, have come to a halt.

Is this the case for the human line? Doesn't the very recent (approximately 100,000 years ago) genesis of the human brain contradict the theory that evolution has lost its creative potential?

In reality, whatever influence hominization may have had on the equilibrium and the future of the biosphere, it was achieved by a very ordinary evolutionary process, but gained enormous importance because of its consequences.

Let us elaborate more clearly. Relatively early in the Tertiary, simians (cynomorphs and anthropomorphs), although greatly developing their brain (to a much lesser extent than man, however), became accomplished climbers, which is their foremost characteristic. Conversely, hominids very early ceased to be arboreal, if they ever were (the shape of the foot is sufficient to put our arboreal ancestry in doubt). They also greatly increased their brain which showed a marked allometry in relation to the other parts of the body (Bauchot, 1972; Portmann, 1972). The brain attains considerable development, particularly through the growth of the neopallium (cerebral hemispheres), which contains billions of neurons with treelike dendrites; the neopallium thus became an organ with new properties (developing in the young child through social stimulation).

Simultaneously, in order to derive bipedal posture from the previous quadrupedal condition, there were changes in the skeleton and musculature (namely, migration of the occipital opening below the skull, reduction of the facial mass, curving of the vertebral column, broadening of the pelvic bones, and attainment of complete plantigrade posture); the anterior limbs no longer assumed a locomotor function, which was now performed exclusively by the posterior limbs.

The qualitative as well as quantitative changes of the brain have been the major theme of hominid evolution; this group has therefore not avoided the specialization to which all zoological secondary branches have been submitted. These changes gave us consciousness and reasoning, and thereby freedom of decision. They also made the loss of automatic behavior possible and harmless, and gave us the ability to adapt to a variety of circumstances without engaging in the sterile direction of organic functional specialization. There is no paradox in saying that the specialization of the hominids has been, thanks to their brain, a counterspecialization. This is why their evolution is unusual, although with regard to anatomical and physiological characteristics (including the nervous system) it does not equal the magnitude of the genesis of a phylum or of a class which creates or deeply modifies the structural plan. In fact, "hominization" is but the evolution of a branching secondary line, whose most remarkable characteristics are the development of the brain and of the hand, which can be compared to monodactyly and hypsodonty in the Equidae.

For the last 100,000 years, *Homo sapiens* (including *praesapiens*) has remained physically stable. Ruler of the earth and of his own evolution, because among all living beings he is the only one capable of assigning an end to his fate, he progresses or regresses at will, hesitating between further "hominization" and reversal to animalism.

Each line, during the course of its evolution, behaves much like a "superorganism" following its own destiny. Several of them follow a three-part scenario, gradually proceeding from one to the next:

A. The "pioneers" of the line, usually few in number, occupy a limited geographical area. Their anatomy shows little specialization but reveals peculiarities which increase in number and importance with the aging of the line. Their evolution is slow, very slow. This is the *period of youth*.

B. This period of setting up, with much hesitation, with trial and error, is followed by a period of great activity; evolution speeds up, rushes headlong; lines radiate, differentiate, specialize, and grow lateral branches. This is the *period of maturity*.

C. Finally, innovations become scarcer; a relative stability replaces rapid variations. Lines die out, others wilt away. This is the *period of senescence*; however, a few lines retain their vigor and still persist.

THE MIOCENE: AN EPOCH OF REFERENCE

Nothing can give us a better idea of the course of evolution during the last 30 million years than the comparison of the Miocene fauna (26 to 12 million years ago) with the modern fauna. This period is far enough in the past to be used as a point of reference.

Climatic and geographical conditions were different from those prevailing today. The seas were hotter and hence more favorable to the development of coral reefs; during the upper Miocene (level of the Pontic), they cooled down and receded. The recession was so great that the Bering isthmus emerged, but the two Americas remained separated. A land which the geologists name the "Caribbean Land" emerged between them.

In Europe, the Germanic gulf extended deeply into the lands lying south of Scandinavia; the North Sea was bounded to the East by Scandinavia, and to the West by the British Isles. Between Vienna and the Caspian Sea, depressions formed large salt water basins having an outlet in the Mediterranean (Pannonian, Pontic, Arabo-Caspian basins), and a deep furrow separated India from the Himalayas.

The climate was generally warmer. Forests of sequoias covered the greater part of northern Europe and Germany. The temperate flora (without palm trees) stretched much farther north; traces of it could be found in Greenland above the 80th parallel north (with fir trees, bald cypresses, poplars, willows, elms, lindens, birches, reeds).

In European regions, however, as early as that period tropical plants gradually started to disappear. Palm trees, camphor trees, and laurels became scarcer or extinct; deciduous trees tended to predominate. In North America, the climate was wet and varied little from north to south; the flora became uniform; most of the genera which composed it (e.g., magnolias, hydrangeas, tulip trees, sassafras, plane trees, sequoias, incense cedar) still thrive today. For our discussion, it should be remembered that the Miocene flora differed from present-day flora more by its geographical distribution than by its composition (there is still a considerable number of living Miocene phanerogams). The higher plants had completed their major transformations.

The marine fauna was very similar to ours, except for its obviously different geographical distribution due to a different climatic distribution. Large Nummulitidae (Foraminifera) were still plentiful; some of them still survive today. Crustceans were practically identical to modern forms, but some groups, such as the polypoid barnacles and Pyrgoma, were more abundant since their habitat was much larger (they live in coral reefs). The bivalve mollusks of the genus Tridacna began appearing in reef formations for the first time (Poland); today, they are located in the coral reefs of the warm Indo-Pacific seas.

A fauna of bivalves developed in the somewhat brackish Pannonic and Arabo-Caspian basins; it was composed of genera which survive to the present day (Dreissena, Congeria, and others).

In European and North American fresh waters, the same genera were found; uncommon forms were rare (for instance, the dextrally coiled snails of the genus *Valenciennesia,* upper Miocene). In the Miocene, the genera *Viviparus* and *Helix* (*sensu lato*) seem to have greatly multiplied, forming numerous species, most of which are still surviving.

Among echinoderms, regular echinoids produced a few new genera (in the families Temnopleuridae, Toxopneustidae, Echinidae), most of which still exist. The irregular echinoids diversified into highly specialized genera: *Monostychia, Mellita,* and *Cleistechinus.* As early as the end of the Eocene, the systematic framework of the echinoids was built, and since then, their tendency toward stability has never stopped increasing.

Lower vertebrates have hardly changed since the beginning of the Miocene (29 to 30 millions years ago). The Viperidae, the earliest representatives of which can be traced back to the Oligocene, developed in the Miocene. Their fossils are usually found in the European continental formations and appear to be very similar to present species.

The Miocene birds increased a rich fauna, elements of which are known to have existed since the Eocene (20 families) or the Oligocene (14 families). The oldest available fossils of the families Struthionidae (*Struthio*), Otidae, Cracidae, Falconidae, Laniidae, Fringillidae, Motacillidae, and Corvidae have been found in sediments of Miocene date.

Several families, mostly belonging to the order Passeriformes, which includes more than half of the living species of birds, presumably differentiated after the Miocene. Numerous species have only been found in fossil form from the Pleistocene (22 families); but the discovery of a fossil does not always indicate the date when its species, genus, or family appeared. A great number of fossils have been dug out from one sediment only to be superceded by one found in an older sediment.

In the case of passeriforms, the numbers are so high (at least 17 families are exclusively Pleistocene) that we have to admit that the evolution of families and genera continued until present times (see Mayr, 1945). During the course of their history the carinates have lost 21 families: Eocene, 10; Oligocene, 6; Miocene, 3; Pleistocene, 2.

The following is an intentionally limited list of surviving genera which have been found in sediments from the Eocene to the upper Miocene (the earliest date only is mentioned):

Albatross	*Diomedea tryidata,* upper Miocene, Australia	
Puffin	*Puffinus calhouni,* upper Miocene, California	
Fulmar	*Fulmarus hammeri,* upper Miocene, California	
Heron	*Ardea piveteaui,* upper Eocene, Plaster of Paris	
Buzzards	*Buteo circoides,* middle Oligocene, western Gobi	
	Buteo pusillus, middle Miocene, Saint Alban, Isère (France)	

Aethia	*Aethia rossmoori,* upper Miocene, California
Sandgrouses	*Pterocles larvatus* and *P. validus,* Phosphorites of Quercy, Lot (France), upper Eocene and lower Oligocene
Owls	*Bubo incertus,* phosphorites of Quercy
	Bubo poirieri, lower Miocene, Saint-Gérand-le-Puy, Allier, France
	Otus wintershofensis, middle Miocene, Bavaria
	Asio henrici, phosphorites of Quercy
	Strix dakota, lower Miocene of South Dakota
	S. brevis, middle Miocene, Bavaria
	Tyto ignota, middle Miocene, Sensan, Gers (France)
	T. sanctialbani and *T. edwardsi,* middle Miocene of la Grive Saint-Alban, Isère (France)
Swifts	*Apus ignotus,* middle Miocene, Saint-Gérand-le-Puy, Allier (France)
	A. gaillardi, middle Miocene, La Grive Saint-Alban, Isère (France)
Swallow	*Collocalia incerta,* middle Miocene. Saint-Gérand-le-Puy, Allier (France)

The above-mentioned species have disappeared, but the genera persist in other very similar species. The only evolution for these and other genera not mentioned in the above list has been speciation.

The mammalian fauna of the Miocene was highly differentiated. The proboscideans, with the mastodons and dinotherians, were fully expanding. Perissodactyls reached their peak (Rhinocerotidae) or evolved rapidly (Equidae). Numerous artiodactyls appeared (Hippopotamidae, Cervidae, Bovidae, Ovidae, Capridae). Creodonts were then represented by only a few species.

Fissiped carnivores had reached or almost reached their peak. The Canidae and Hyaenidae came into existence and large-sized Felidae were greatly developing. Some of them with saberlike upper canines, the genus *Machairodus* of Eurasia, replaced the Oligocene genera of similar habitus, which were completely wiped out. The pinnipeds appeared in the lower Miocene with the Otariidae, the genera differing from living ones. Fossil seals, including *Phoca,* have been found in sediments of the upper Miocene.

In South America, the notoungulates and their allies, the litopterns, differentiated into several categories. The xenarthid edentates continued evolutionary trends already well underway in the Oligocene or Eocene; new branches were grafted on the main branches. They reached their peak in the Pliocene and Pleistocene, at the end of which their decline accelerated; several families, the Mylodontidae, Megalonychidae, Megatheriidae, and Glyptodontidae, became extinct shortly before the Quaternary, or at its beginning. The armadillos (Dasypodidae) were the only family to withstand successfully the test of time. The first anteaters (Myrmecophagidae) can only be truly recognized from the Pliocene of Argentina. The origin of sloths (Bradypodidae) is still unknown; they belong to a suborder, the Tardigrada, which was very successful during the Tertiary era in South America, and gave rise to several lines. Because of the lack of fossil evidence, the origin of the Bradypodidae is arbitrarily placed among paragravigrades.

The causes of the abrupt extinction of glyptodonts and gravigrades at the end of the Pleistocene are unknown. Since changes in climatic conditions did not affect the entire area in which they were distributed (South America, California, and New Mexico), it is most likely that such conditions were not the sole agents of this extinction which affected families at various levels of specialization.

Pangolins, which appeared in the Oligocene, lived in a much greater geographical area than they do at present (Africa, tropical Asia) and could be found in Europe; today they have reached their level of stabilization and are becoming scarce. Lagomorphs presumably appeared between the Paleocene and the upper Eocene; several genera were already quite representative of the order; they were continued in the Oligocene, for example, by the genus *Palaeolagus* (Oligocene of North America). They are also found in Miocene sediments of North America, which seems to have been their birthplace. Their appearance in Europe is fairly recent; fossils of rabbits (*Oryctolagus cuniculus*) have been found in Villafranchian sediments.

Typical or glirine rodents, form a very full order, showing a great number of genera and species, many of which have a strong endemic character (neoendemism). Although it has been clarified by Schaub, their classification, based mainly upon the characteristics of the teeth, is still not entirely satisfactory. It would seem that the evolution of the order cannot be established with any degree of accuracy or certainty.

Eighteen families appeared in the Eocene–Oligocene, six in the Miocene, five in the Pliocene, and three in the Pleistocene. These statistics, based on a limited number of fossils, can only be considered as suggestive, but they show that during the Tertiary era, the glirines never stopped evolving and produced new families up to the Quaternary. A considerable number of genera have been produced: approximately 350 exist today (1952 estimate) and 290 are fossils (1958 estimate), to which should be added the numerous fossils destroyed or still to be found. During the course of evolution, the typical "glire" structure has been preserved despite its variations. It can be found both in the capybara and the harvest mouse, in the beaver and the squirrel, and in the meadow mouse and the scaletail. In the Miocene, the evolution of the glirines had already slowed down but continued nonetheless during the succeeding periods.

The Insectivora reached their peak in the Oligocene. In the Miocene, however, this key order, from which the Chiroptera, Dermoptera, and primates were derived, had already lost most of its species. The Solenodontidae still survived; but today only three species can be found, in minute number, in Cuba and Haiti. Shrewmice, moles (well adapted to an underground existence), and hedgehogs were clearly differentiated. Many genera and the family Dimylidae (showing numerous hedgehog characteristics), which are now extinct, belonged to this order. No fossil belonging to the Macroscelidae has yet been clearly identified. *Anagale,* a fossil genus of tree shrews, has been found in an Oligocene sediment in Mongolia; there is no known vestige of the tree shrews in the Miocene.

It appears clearly that the Insectivora have not given rise to any innovation since the Miocene: their evolution has been limited to the elaboration of several species and a number of genera. However, this group, in which highly specialized species (e.g., muskrats, moles) coexist with species of a general structure, continues to thrive. The shrewmice (*sensu latissimo*) and the moles easily adapt to the proximity of man.

Our knowledge of Miocene primates is very incomplete, as we only possess a few lemuromorph fossils and no tarsier fossils. They did exist, since they left some remains in Eocene and Pleistocene sediments. The Lorisidae are identified from the Miocene of Kenya (*Indraloris* and *Propotto*). They were very close to the modern species.

Simians have already been discussed; refer to page 63.

No fossil dug out from Miocene sediments has yet been definitely identified as belonging to a hominid. There is some doubt regarding *Ramapithecus* (= *Kenyapithecus*), which was found in the Sivalik mountains (India) and in Kenya (14 million years old), but the available jaw and tooth fragments do not allow definite classification. Hominid or pongid ape? No one knows.

In the Miocene, the evolution of simians and anthropoids was practically completed, while the evolution of man still had a long way to go.

THE DAMPENING OF EVOLUTION

From the facts already discussed, one notices that the "maneuvering space" of evolution has never stopped decreasing. The genesis of the phyla stopped in the Ordovician; of the classes, in the Jurassic; of the orders, in the Paleocene–Eocene. After the Eocene, the evolutionary "sap" still flowed through a few orders, since mammals and birds continued to specialize in various directions and invaded all the terrestrial and marine biotopes previously occupied by reptiles.

The extent of evolutionary novelties gradually changed. They no longer affected the structural plan but only involved details.* The only form which evolution took was speciation: in insects since the Oligocene, in mollusks since the Miocene, in birds and simians since the Pliocene, and in some glirines and hominids since the Holocene; *Homo sapiens*, the last in line, is probably 100,000 years old.

Evolution has not only slowed down, but with the aging of the biosphere, it has also decreased in scope and in extent. We are certain that it does not operate today as it did in the remote past. Something has changed. It is of the utmost importance to determine what has changed; this should shed light upon the internal mechanisms of the phenomena. The structural plans no longer undergo complete reorganization; novelties are no longer plentiful. Evolution, after its last enormous effort to form the mammalian orders and man, seems to be out of breath and drowsing off. I find this metaphor a good description of the present state of evolutionary phenomena.

Has evolution always resulted from the same causes? Have the mechanisms always been the same? Since the fundamental functions and the chromosomal cycles of cells have apparently remained unchanged in multicellular animals, we are inclined to believe that this is true. However, the varying rates, the various halts, and the inability to produce new structural plans do not support the concept of uniform or constant evolutionary mechanisms. One may also assume that the "effectors" have lost or gained a factor which prevents them from evolving as in the past.

The period of great fecundity is over: present biological evolution appears as a weakened process, declining or near its end. Aren't we witnessing the remains of an immense phenomenon close to extinction? Aren't the small variations which are being recorded everywhere the tail end, the last oscillations of the evolutionary movement? Aren't our plants, our animals lacking some mechanisms which were present in the early flora and fauna?

It has often been noted that, despite the presence of all the presumably efficient causes, evolution still stops. Vandel (1972) has recently

---

*Meadow voles of the genus *Microtus* are an example of recent rapid microevolution; they are derived from a stem species, *Allophalomys pliocaenicus*, the remains of which have been found in sediments dating back to the end of the Villafranchian (less than a million years old). Small variations, arbitrarily and subjectively classified into various categories, have gradually transformed the genus *Allophalomys* into *Microtus*, which is one of the genera with the most abundant living species (more than 90 species and 240 subspecies). The genus *Microtus* is palearctic and can be found in Mexico and North America; meadows are its favorite habitat (Chaline, 1972, 1973).

elaborated a very good example of this. The two species of woodlouse of the genus *Australoniscus*, *A. alticolus* in Nepal and *A. springetti* in western Australia, have been separated, because of the splitting of the Gond-wana continent and because of continental drift, since the beginning of the Cretaceous (i.e., approximately 140 to 135 million years ago). They differ by a minor characteristic; "the endopodite end of the first male pleopod is different... it is straight in *springetti,* bent into a hook in *alticolus.*"

Thus, in 140 million years, neither segregation nor mutations (there certainly have been some), nor selection operating in different environments, has provoked any change in these crustaceans. The cause of their stability must therefore lie in their inner structure. Does it lack the necessary means to encode new information in its gene pool, or is a given mechanism hindering this process? No one can tell.

In Chapter VII an attempt will be made to answer such questions by reference to the recent data provided by molecular biology. It can, however, already be concluded from this survey that, today, evolution is no longer what it once was. This highly significant fact has been little considered by biologists, who should not look exclusively for the mechanism of evolution, but should also reveal the causes which stopped the creation of new types and which altered the speed of the process.

The curve of evolution demonstrates that it is the result of a series of irreversible historical phenomena. Evolution does not repeat itself in identical sequences, nor does it reverse itself. It is really a history which resembles that of a nation or of a race. Thus, hominid evolution gradually passes from the field of paleontology to that of true history.

THE GENEALOGICAL TREE OF THE ANIMAL KINGDOM

In order to clarify certain problems, let us first consider the family trees established by Cuénot (1952) and myself (Grassé, 1969, 1970), based on data derived from paleontology, comparative anatomy, and biochemistry. Since no fossil belonging to the direct and close ancestry of the phyla is known, any family tree is highly conjectural.

All vertebrate fossils are unquestionably related to existing classes; none of them can be considered as the possible ancestor of either the agnathans or the jawed fishes (see Chapter I). The evolutionary rise toward the vertebrate type has been the subject of numerous theories (Geoffroy Saint-Hilaire, 1818; Patten, 1890; Mastermann, 1897; Gaskell, 1908; Garstang, 1929; Romer, 1955; Jarvik, 1960; and a few others); they are mostly speculative and need not concern us here. Crossopterygian fishes, which have been recently and thoroughly studied, may have been the synthetic group from which tetrapod vertebrates evolved.

The actinistian crossopterygians remained unchanged from the upper Devonian to the Jurassic. They were much too conservative and specialized, judging from the modern *Latimeria*, to have played a part in the genesis of tetrapods, the "conquerors" of terrestrial environments. This role was presumably played by the rhipidistian crossopterygians which, like the tetrapods, possessed internal nares (choanae).

The earliest known amphibians, the ichthyostegalian stegocephalians, appeared in the upper Devonian (Old Red Limestone of eastern Greenland); they are unquestionably similar to the rhipidistians, especially to the Osteolepiformes, but they show characteristic tetrapod limbs (called autopodia),* attached to the trunk by characteristic bony girdles. The intermediate links between the crossopterygian fin and the tetrapod limb are missing. Anatomists have been trying to identify in the fin the bones which formed the autopodium. Their task is not an easy one.

Each lobed pectoral fin of a crossopterygian is composed of a proximal bony element, homologous to the tetrapod humerus, followed by two bony elements to which are attached rows of ossicles. The fin articulates with a girdle consisting of five bones, different from that of other fishes. The anterior, scapular girdle is attached to the head on either side by two bones, whereas that of amphibians is free.

In spite of the wide differences between the lobed fin and the autopodium, it is most likely that the latter was derived from the former. *Ichthyostega* bore a heterocercal caudal fin, constructed somewhat like that of crossopterygians. The skull arches of the osteolopiforms and ichthyostegalians do not have the same structure but the cheek bones are practically the same in both forms. Thus, the ichthyostegalians share too many characters with the crossopterygians for the latter not to belong to their ancestry; the crossopterygians are not, however, direct ancestors of stegocephalians, but are apparently very close to them. *Ichthyostega, Ichthystegopsis,* and *Acanthostega* have been considered a side branch of the crossopterygian–amphibian stock which, after a rapid evolution, withered away without any offspring. In order to measure fully the relationships between these archaic groups, one would have to know which specific transformations of the respiratory and circulatory

---

*The characteristic limb of walking Vertebrates (tetrapods) which is composed of successive articulating segments; forelimb: upper arm (humerus), forearm (radius and ulna), wrist (carous), palm (metacarpus), and digits; hind limb: thigh (femur), lower leg (tibia and fibula), ankle (tarsus), sole of the foot (metatarsus), and toes. The basic number of digits and toes is five per limb.

systems made terrestrial life possible, but no known facts provide any clues.

The transition from amphibians to reptiles was also accomplished through unspecialized forms. It was so gradual that certain fossils, such as *Seymouria,* were initially mistaken for primitive reptiles; they are now classified as the Seymouriamorpha and placed among the labyrinthodont stegocephalian amphibians.* Their skeleton shows a variety of amphibian and reptilian characters: their general body form is quite common, similar to that of a triton, or of a lizard. In the skull, reptilian features are not abundant (lachrymal extending from the orbit to the external narest; however, the vertebrae of *Seymouria* and of cotylosaurian reptiles are almost identical. Reptiles possess two or more sacral ribs and amphibians only one, but *Seymouria* has two. Its limbs and girdles show several clearly reptilian characteristics. The order Seymouriamorpha possesses few unique characteristics (shape and position of the otic notch, formation of the atlas and axis complex, presence of two sacral ribs) and none of its representatives shows any sign of a physiological or ethological specialization. We must note that the archaic form *Seymouria* has strong affinities with the most primitive reptiles, the Captorhinomorpha. We have already described in some detail the genesis of mammals from theromorph reptiles, which are remarkable for their archaic and general structure.

Our evidence on birds is less complete, although there is the outstanding *Archaeopteryx,* which exhibits a real mixture of reptilian and birdlike characters. Its ancestors are unknown; consequently, the precise origin of birds is still to be discovered.

THE PARENT FORMS AND THE CREATION OF NOVELTIES

We now have sufficient knowledge to affirm that *derivation of one type of organization from another never occurs through specialized types.* The great evolution was achieved from archaic forms to archaic forms; each form, by its structural plan, clearly belongs to a given systematic unit, but

---

*The seymouriamorphs are undoubtedly amphibians. On the skull of some specimens of *Seymouria,* traces of sensory canals (characteristic of fishes and amphibians) have been observed.

†This is a very archaic feature since it is present in *Ichthyostega;* it is also found in cotylosaurian reptiles.

retains a structure of the general type. These forms are so little specialized that, considered singly, they hardly reveal anything about their development. Their evolutionary interest and significance is only apparent when they are included in the line or lines from which they derive. They are the "parents" which gave rise to the phyla; the latter, by giving way to specialization, build a given morphological type, or *idiomorphon*.

Once engaged in a given evolutionary pathway, a given line cannot escape from it; it can at the most increase its specialization. The branches do not produce any really new creation. They can be compared to the masses of cells which in the course of ontogenesis, after undergoing the influence of an "organizing agent," can only form the organ for which they were prepared.

The parent forms correspond to the trunk, or axis, of classically conceived family trees. They possess the creative potentialities; they are comparable to a rhizome from which shoots grow here and there.

Creative evolution originates in parent forms; if these are absent, new types of organization never appear. For instance, reptiles (whether *Diadectes* was a reptile or an amphibian is of little importance here) were evolved from archaic stegocephalian amphibians of the seymouriamorph unspecialized type.

The history of mammals reveals that they derived from very archaic reptiles, the captorhinomorphs; these are closely related to the stock, even are the very stock, from which also arose, at about the same time, all reptilian orders. Very early parent forms sent out side branches which correspond in number to the subtypes, and on which are attached second and even third rank side branches.

Toward the end of the Cretaceous or the beginning of the Paleocene, Insectivores, Chiroptera, and primates were evolved from the same stock as primitive mammals. Later, tarsiers, lemurs, simians, and probably hominids came to be separated from a common stock. Direct phyletic links are unlikely between the first three suborders. Further, I believe—but this is a personal opinion—that the direct derivation of hominids from the common stock of all primates is more probable than derivation from the simian line.

The picture of the great evolution is not exactly that of a tree but rather that of a rather short stem which, at each node, sends out a whorl of unequal lateral branches (see Fig. 5). The thicker branches, which represent the classes, send out secondary twigs (orders), tertiary twigs (suborders), and quaternary twigs (superfamilies or families). The leaves represent the species.

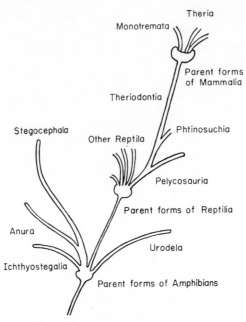

Fig. 5 Derivation of amphibians, reptiles, and mammals. The above is an example of a genealogical tree bearing whorls of evolutionary "twigs." The "parent" forms of amphibians are unspecialized rhipidistian crossopterygian fishes; the ancestral, "parent" forms of reptiles are seymouriamorph stegocephalians and captorhinomorphs; the "parent" forms of mammals are pelycosaurs and theriodonts.

PANCHRONIC FORMS AND ARRESTED EVOLUTION

In some cases, the pace of evolution is so slow that one can say it is arrested. Many species have not evolved since their origin, or have hardly done so, and are the surviving forms of extinct groups. They are known as *relicts*. The following are some classical examples.

Lingulas (*Lingula*) and Cranias (*Crania*), Ordovician and Carboniferous brachipods, are present in our seas.

Monoplacophoran mollusks lived during the Cambrian–Silurian. Two species of the genus *Neopilina* have recently been discovered; the first lives in the Pacific Ocean off the western coast of Mexico at a depth of 3500 meters; the second much deeper, between 5600 and 6350 meters.

Gastropod mollusks comprise numerous, very ancient genera: *Viviparus* (corresponding to the modern *Paludina*) left shells in lower Carboniferous sediments; *Potamides*, in Permian sediments; *Valvata* appeared in the middle Jurassic, trochids (*Trochus* and other genera) in the Silurian, fissurellids (*Fissurella*) and pleurotomariids (*Pleurotomaria*) in the Jurassic. Present lamellibranchs include numerous relict forms: *Nucula* (Devonian), *Leda* (Silurian), *Arca* (Triassic), *Limopsis* (Triassic), *Modiola* (Devonian), *Lithodomus* (Carboniferous), *Mytilus* (middle Triassic), *Pinna* (middle Triassic), *Avicula* (Devonian), *Chlamys* (Triassic), *Spondilus* (Permian), *Anomia* (Jurassic), *Astarte* (middle Triassic), *Isocardia* (Jurassic), *Cardium* (Triassic), *Teredo* (Jurassic), etc.

Several general of nautilids are present from the upper Cambrian; the only surviving relict forms, *Nautilus* (three species), are directly related to the Nautilida (first forms dating from the Mississippian: 320 million years ago); the shells of *Nautilus* (*sensu stricto*) have been found in Eocene sediments.

Syncaridan crustaceans, known in fossil form from the Carboniferous to the Permian, include several genera (*Koonunga, Anaspida, Paranaspida, Micraspida, Bathynella,* and others) which live in underground waters of the whole world, except for North America.

Among insects, the odonates *Epiophlebia superstes* (Japan) and *E. laidlawi* (Himalayas) are the surviving forms of the Anisozygoptera which were abundant from the Triassic to the Liassic.

Tricholepidion gertschi, a thysanuran living in northern California forests (vestige of the mesotypic forest of the Tertiary), is the only surviving species of the lepidotrichinate thysanurans caught in the amber of the Baltic. *Mastotermes darwiniensis* (northern Australia) is the only representative of the Mastotermitidae which, in the Eocene, lived in Europe.

Among fishes, relicts are abundant. Coelacanths, which have already been mentioned, live in the Indian Ocean, off the Comoro Islands, in waters 150 to 800 meters deep; they are very similar to *Coelacanthus,* the remains of which are found in Carboniferous (Ohio, Scotland, England), Permian (Europe), and lower Triassic (Madagascar) sediments. It may be presumed that *Latimeria* has not evolved at all for the last 200 million years!

The remains of several living selachian fishes have been found in fossilized form in upper Triassic (*Pristiurus,* for example), Cretaceous (*Sphyrna*), and Eocene (*Raja, Squalus*) sediments. These fishes have not changed since then. The family Chimaeridae (holocephalian fishes) existed already in the Jurassic (Kimmeridgian of Bavaria, Bathonian).

The sphenodon or tuatara (*Sphenodon punctatus*), found today in the small islands of the Cook Strait (New Zealand), is the only rhynchocephalian reptile (Triassic, peak in the Jurassic) which has survived.

Mammals also have relicts: the opossums (*Didelphis*), for example, are extremely close, except for a few details in the dentition and the skeleton, to *Eodelphis* of the upper Cretaceous (Alberta, Canada) (60 to 70 million years ago).

The hyracoids have hardly shown any change since the lower Oligocene: *Sagatherium* and *Pachyhyrax,* found in the Fayum beds (Egypt), are close to the African *Procavia.* Insectivores of the genus *Sorex* were present from the lower Oligocene.

This list is only a small sample of the considerable number of relict animals which have been reported (see Jeannel, 1944; Delamare-Deboutteville and Botosanéanu, 1970).

The existence of panchronic species raises several questions: What causes slowed down their evolution and brought it to a standstill? The answer is all the harder to find since the panchronic species can be divided into two different groups: those that have a general structure close to the original stock, e.g., the opossum (*Didelphis*), the selachian fishes, that adopt neither a specific habitat nor a specific way of life; and those that have a mixed (half archaic, half specialized) structure, e.g., *Amphioxus,* coelacanths, tarsiers, *Orycteropus,* all of which have a highly specialized habitat or way of life.

We do not know why so many sharks (some of them for more than 100 million years) and opossums (*Didelphis*) (for 70 millions years) have remained perfectly stable. Yet the opossums, which live in such widely different environments as damp forests, savannas, subdesert areas, and the edges of town, are subject to conditions theoretically favorable to evolution. Some of their species, *Didelphis virginiana* (North America) and *D. azarae* (South America) are widely distributed and mutate extensively. These are quite healthy relics, but they refuse to evolve.

Some panchronic species only persist because of the favorable conditions they find in a few rare ecological niches, their refuges. The small islands of the Cook Strait (New Zealand), where the last rhynchocephalian (*Sphenodon punctatus*) lives, are the refuge in which this species finds the environment suited to its physiological peculiarities: very low metabolism, very slow growth (average growing time, 50 years), sexual maturity reached at approximately age 20, longevity thought to be about a century (Dawbin, 1962).

The fact that specialization stops evolution seems likely in many cases: *Amphioxus,* which shows chordate organization in a simple form and foreshadows vertebrates, lies burrowed at the bottom of the seas, with its anterior pharyngeal and brachial end protruding and forming a net used to catch plankton. Microphagy (feeding on microscopic animals, bacteria, algae, or minute organic particles floating in the water) requires the possession of complex filtering devices. The biggest and most specialized filtering device can be found in the mysticete Cetacea, the product of a highly elaborate evolution.

Orycteropus, a mammal showing a very archaic mammalian structure, is not unrelated to the Condylarthra, which lived from the Paleocene to the Eocene. It feeds exclusively on termites: its reduced dentition is of an aberrant type (tubulidont); the sticky saliva of its tongue (abundantly secreted by enormous salivary glands) picks up insects. Its claws and snout enable it to burrow quickly and efficiently. The other parts of the body have remained simple and archaic; it is as if its specialization had caused the animal to retain its archetypal structure.

Tarsiers, which are primates, adopted very early in the course of their history arboreal habits, an insectivorous diet, and a nocturnal pattern of activity. This way of life is in keeping with a hypertrophy of the eyes, which is so marked that the volume of a single orbit can reach the size of the brain cavity. The elongate and slender toes probe the cracks in the bark in order to gather insects. To the adaptations of day-blindness and insectivory was added that of the arboreal leaper: the hind legs, much stronger than the forelimbs, have an enormously elongated tarsus. Specialized to the extreme, tarsiers have split up into three species and 12 subspecies which show slight differences; this proves that specialization is not creative evolution.

PERSISTENCE OF EVOLUTION

The fact that a group is very ancient does not, however, mean that evolution is arrested. The echinoderms illustrate this fact. They appeared very early and some of them, the Cystoidea, have left fossilized remains since the Cambrian. They reached their peak in the Ordovician and disappeared in the Devonian. The Blastoidea lived from the Ordovician to the Permian, the Edrioasteroidea from the Cambrian to the Carboniferous. These two classes are totally extinct.

The oldest forms of the Crinoidea date from the lower Ordovician; they still have representatives which all belong to the subclass Articulata, of Triassic age.

The Stelleroidea, which are the same age as the Crinoidea, are still widely represented in all seas, but their evolution has been very slow. The genus *Astropecten*, which is fossilized in Jurassic sediments (Bajocian), can still be commonly found in present oceans, as well as the genus *Echinaster*, which dates from the Cretaceous. A great many existing asteroids seem to have evolved recently.

The history of the Echinoidea is particularly well known; it began in the Ordovician and has been unfolding, at variable speeds, ever since. Cidarids appeared in the upper Triassic although they do not lack relatively recent genera (Eocene). Some families, the Strongylocentrotidae and Echinometridae among regular forms, and Spatanqidae, Lovenidae, etc., among irregular forms, are most Miocene.

A great number of existing genera have been found in fossil form. A few of them are *Schizaster* (Eocene), *Parabrissus* (Eocene), *Spatangus* (Eocene), *Brissus* (Eocene), *Hemiaster* (Aptian), *Cassidulus* (Senonian), *Oligopodia* (Eocene), *Echinolampas* (Eocene), *Clypeaster* (Eocene), *Echinocyamus* (Senonian), *Laganum* (Eocene), *Dendraster* (Pliocene), *Astriclypeus* (Miocene), *Rotula* (upper Pliocene), *Stereocidaris* (Cretaceous), *Coelopleurus* (Eocene), *Microcyphus* (Miocene), *Erbechinus* (Pliocene), *Sphaerechinus* (Pliocene), *Echinus* (Pliocene), *Psammechinus* (Neocomian), and *Parasalenia* (Miocene).

In sum, the phylum Echinodermata, which in the past was divided into eight subtypes or classes (Cystoidea, Blastoidea, Edrioasteroidea, Crinoidea, Holothuroidea, Stelleroidea, Ophiuroidea, Echinoidea), has lost three lines but has nonetheless continued to be present in all the seas and at all depths. The Cystoidea were superseded by the Blastoidea, then by the Crinoidea. The Echinoidea and the Stelleroidea have been similarly sucessful from the Ordovician onward; as for the Holothuriidae (the oldest fossil of which has been found in lower Devonian sediments), they do not show any signs of senescence. Thus, despite their antiquity the echinoderms have been constantly evolving at various rates depending on the period and the class to which they belong.

The existence of relict forms (i.e., those remaining unchanged for a very long period of time and surviving an otherwise extinct group) is only a particular aspect of arrested or retarded evolution, the phenomenon which explains the persistence of the lower phylogenetic branches. Bacteria, protozoans, sponges, and cnidarians (hydroids, jellyfishes, corals) are still plentiful; they appear under the same form as in much earlier periods and still produce mutations.

The modern protozoans, taken as a whole, differ, but only slightly, from the ancient ones. They resemble very closely the ancestral forms which were floating in the Paleozoic oceans. Their classes, orders, and families claim a remote antiquity. Within these categories, however, evolution was highly diversified; the genera of Radiolaria and Foraminifera became extremely abundant without transgressing the unicellular frame. The Foraminifera, which are still plentiful, reached their peak in the Tertiary seas, becoming the characteristic fossils of marine sediments; geologists call the period of their greatest development the Nummulitic period (*Nummulites, sensu lato,* are Foraminifera with a complex skeleton, appearing in the Jurassic and now represented by several genera).

It is almost unnecessary to point out that protozoans (in the course of their evolution) have suffered losses; several groups, now extinct, are proof of this; one might mention the archaeomonadids among the Chrysomonadidae, the Vallacertidae among silicoflagellates, several families of dinoflagellates, ebriedians (excluding the genera *Ebria* and *Hermesinum*), and various foraminiferous families.

Radiolaria have left fossilized remains in the Precambrian sediments of Australia. The two main groups, the Spherellaria and Nassellaria, existed already in the Paleozoic, but had forms slightly different from the existing ones. The evolution of certain lines is known with some degree of accuracy; it lasted until the Tertiary. The phylum Porifera, which originated in the Precambrian, has evolved little since the Paleozoic era. Moret (1952), an authority on fossil species, wrote: "As regards evolutionary phenomena, they are almost nonexistent in sponges," which exhibit "a remarkable inability to undergo transformations in the course of time" (it has been said that sponges are a permanent challenge to evolution). To exemplify the previous statement, one might look at the most complex forms (skeleton and arrangement of filtering chambers) of the hexactinellids; as early as the Silurian they had representatives, the Lyssacina, whose body organization is very similar to that of the living species. Despite their antiquity, the Porifera are present in all faunas, in all waters, from coastal to abyssal.

Brachiopods form a phylum or subphylum, whose evolution, although never of major magnitude, produced several superfamilies which have prevailed in succession. It is a declining group: 1300 fossil genera have been recorded, but it probably included many more genera. Its history is complicated: it seemingly went through two successive peaks, one in the Paleozoic (Ordovician–Devonian) and a more recent one, in the Mesozoic. Its current 68 genera, which include several venerable relicts (*Lingula, Crania*) hold a modest place in ocean life. Its evolution affected details without altering the basic structural plan.

The Collembola, an order of insects, are an example of a very ancient, surviving group. As early as the Devonian they were fully individualized (see the case of *Rhyniella praecursor*); they are highly prosperous today and colonize most environments, especially moist ones: soil, leafmold, humus, rotten wood, stream banks, mountain areas, decaying vegetable substances, living vegetation, caves, and termitaries. Authorities estimate that there are about 100,000 species of Collembola!

But this exuberant production of species contrasts sharply with the extreme monotony of the structural plans, for we identify only two suborders, the Arthropleona and the Symphypleona. Genera and species are distinguished by slight details, revealed only through the microscope. The uniformity of the structural plans does not impede genetic and specific diversity. Infusorians, beetles and birds, to mention only very distinct animals, also provide conclusive examples.

Despite their antiquity, the Collembola exhibit a morphological as well as a physiological adaptability, which results in intense production of ecomorphoses. Several of them, which have been carefully studied by Cassagneau, are somewhat similar to the variations occurring in plants transferred from a given environment to another. They also give rise to numerous mutations from which small subspecies apparently derive. These have been described by systematists and are reminiscent of the "jordanons" in *Erophila*.

Is it not clear, then, from the above that neither mutagenesis nor the plasticity of forms alters the stability of the order?

SPECIATION—THE ONLY PRESENT FORM OF EVOLUTION

The evolution of all zoological groups was initially highly productive, then slowed down and is now restricted to the creation of new species. It seems to us that evolution is not more productive in plants than in animals. Creative stages ended long ago, except in birds and mammals which became individualized at the beginning of the Tertiary and specialized during that era. Now their evolution is also confined to speciation.

The invertebrates, all very ancient, have not changed their structural plan for a very long time. The entomological fauna found in fossil resins, especially in the Baltic amber (Oligocene, approximately 33–26 million years old), differs very little from the modern fauna; its exact equivalent is found in tropical or subtropical climates. Kelner-Pillault's studies (1969) on bees embedded in amber reveal one of the strongest cases of evolution undergone by insects during the last 30 million years. These hymenopterans have short mouth parts and pollen brushes with few hairs. Apart from these peculiarities, they are very close to the four existing species of *Apis*.

We are almost certain that, at present, evolution is restricted to episodes of the genesis of a few new species. Yet, there still exist a number of groups from which we could have expected more, especially among sponges, Hydraria, Platyhelminthes, Polychaeta (annelids), Polyplacophora (mollusks), sharks, urodeles, several mammals (e.g., Didelphidae, Soricidae, Tupaiidae). They are archaic types with a general anatomy which could evolve into more specialized lines. In the case of mammals, we refer mainly to two families of Insectivora, the Soricidae and the Tupaiidae. The shrews, for instance, are apparently close to the original insectivore stock. As early as the upper Eocene, they were present in Western Europe as the genus *Saturnina,* and in the lower Oligocene, in North America, as the genus *Domnina* which is similar to *Sorex* and *Blarina.* We are certain that shrews have evolved very little since remote times, producing species which differ only slightly from the original stock. This conclusion does not say much for the evolutionary chances of the Soricidae.

The Tupaiidae, wrongly classified among primates, are semiarboreal, semiterrestrial Insectivora, which are legitimately considered as being arrested in an archaic state. They are known in fossil form as the genus *Anagale* (lower Oligocene of Mongolia), which is only slightly more archaic than the Tupaiidae. *Tupaia* comprises several species and subspecies in Indonesia (including the Celebes Islands). They could apparently still evolve in various directions.

Among American marsupials, some genera such as opossums (*Didelphis*) and the Caenolestidae escaped any form of specialization. *Marmosa, Metachirops,* and *Monodelphis,* although tree dwellers, do not seem to show any anatomical features characteristic of their way of life. Despite their primitive generalized structure, they remain "unchanged" and are not heading toward elaboration of a different form.

The monotremes, in spite of their very archaic features, are so specialized that any innovation seems impossible. Among the least changed eutherian mammals we find the Ursidae, whose reduced sectorial dentition allows an omnivorous diet; their evolution seems nevertheless to have considerably slowed down. Although the date of their origin is still undefined, we know from the genus *Ursavus* that their morphology was already well advanced in the Burdigalian–Pontian although they were still small. At the beginning of the Pliocene the genus *Ursus* was individualized, giving rise to numerous species, a number of which have become extinct; the great cave bear (*Ursus spelaeus*) disappeared only in the Magdalenian, after having been worshipped by Neanderthal man. Therefore, there is no reason to expect great evolutionary prospects from the Ursidae.

The case of typical rodents is different. Even paleontology does not enable us to distinguish fully the evolutionary trends which have affected this order; the affinities between the various families are poorly known. Several mammalogists have claimed, without conclusive evidence, that some are undergoing an "extensive evolutionary process," especially that of the Rattidae. Of all mammalian orders, this order is the richest in genera and species, and some genera include dozens of species (*Microtus* 90, *Rattus* 270)! Within this order, extensive speciation seems to have taken place and recent work on systematics and the chromosomes of North American *Microtus* and *Peromyscus* and of African mice of the genus *Leggada* suggest that it is still taking place, but that it is very limited in scope. Cosmopolitan species, such as *Rattus rattus, Rattus norvegicus,* and *Mus musculus,* live in environments which differ from one town or region to another; one cannot fail to note that under these conditions, supposedly ideal for rapid and varied evolution, such species, although producing a great many mutations, exhibit a marked stability.

Some rodents from Madagascar (the Nesomyinae, a subfamily of doubtful homogeneity), from New Guinea (*Mayermys, Crossomys, Microhydromys,* etc), from Australia (*Hydromys*), may be of a fairly recent origin; some new species may even be appearing in these big islands.

A critical analysis of zoological groups, including fossil and recent species, reveals that most of them "froze" in their present state a very long time ago, and that the less ancient ones are going through a lull. In the last 10 to 400 million years, all of them have exhibited only very slight variations.

Biologists find it hard to admit that, in their basic structure, present living beings differ at all from those of the past. To begin with, such a supposition seems contrary to the scientific spirit. But facts are facts; no new broad organizational plan has appeared for several hundred million years, and for an equally long time numerous species, animal as well as plant, have ceased evolving.

We have said that evolution in the present is difficult, if not extremely difficult, to observe. Some biologists maintain that they can not only observe it but also describe it in action; the facts that they describe, however, either have nothing to do with evolution or are insignificant.

At best, present evolutionary phenomena are simply slight changes of genotypes within populations, or substitution of an allele by a new one. For example, the mutant carbonaria of the birch moth, *Biston betularia,* replaces the regular butterfly in polluted industrial areas (Haldane, 1956; Ford, 1971). It is worth mentioning the neoendemism of air-breathing mollusks of the genus *Achatinella* peculiar to the Hawaiian islands (Welch, 1938), and of the genus *Partula* peculiar to Polynesia (Crampton, 1916, 1925, 1932; Baily, 1956; Murray and Clarke, 1966). The "microspecies" (jordanons) of *Erophila verna,* of the family Cruciferal, are mutants which replace each other depending on the location. Jordan (1864), having noticed their stability, used it as evidence that the pulverization of the species into subunits cannot be regarded as an evolutionary trend. In this respect he is right, since jordanons do not vary and do succeed each other in a preferential direction. In fact, the Linnaean species *Erophila verna* maintains its unity and its validity in spite of the hundreds of varieties under which it can be found. But Jordan was wrong in drawing the conclusion, from the stability of the jordanons, that evolution did not exist. From the observed facts he drew a conclusion which they did not imply.

Minute differentiations of the species, resulting possibly from the adjustment of the genotype to local or other conditions, do not lie outside the framework of the species; these differentiations alter the structure and functions of the species for adaptive "purposes." In this respect, the case of the human species is worth considering.

Man is one of the most cosmopolitan terrestrial animals; he lives in all kinds of climates. He underwent several thousand types of mutations, judging from the number of alleles reported in the various human populations presently comprising three billion individuals, all showing different genotypes (except for homozygous twins). The potential supply of mutants for selection is thus very abundant. What has happened, then? Nothing important or even noticeable. The last anatomical characteristic acquired by man was the chin (30,000 to 40,000 years ago and probably more if we take into account *praesapiens*).

In a recent study, Vallois (1972) reviewed the various races of the European upper Paleolithic; this eminent anthropologist reached the conclusion that only two races really existed: the Cro-Magnon man, whose eponym skeletal remains date from the Aurignacian, and the Chancelade man, whose eponym skeletal remains date from the Magdalenian. These two fairly well-characterized races closely resemble the existing Europoids. Thus, since the Aurignacian, the men who lived and still live in western Europe have at the most added or lost a few minute details of their anatomy. They have *not* undergone *any* significant variation. Vallois writes about the skeleton: "It belongs entirely to the Cro-Magnon type which is that of all French Magdalenians and results from a progressive change of earlier classical Cro-Magnons." It is likely that the present Europoids are derived from the Cro-Magnon and the Chancellade Man of the upper Paleolithic.

Mutations do differentiate individuals, but the human species, despite the magnitude of its population and the diversity of its habitats, both of which are conditions favorable for the evolution of the human species, exhibits anatomical and physiological stability.

In wealthy western societies natural selection is thwarted by medical care, good hygiene, and abundant food, but it was not always so. Today in underdeveloped countries, where birth and death rates are equally high (tropical Africa, Amazonia, Pakistan, India, Patagonia, some Polynesian islands), natural selection can exert its pressure freely; yet the human type hardly changes. In the population of the Yucatan, which since the Spanish conquest has been subjected to terrible vicissitudes, one can find Mayan men and women who are the exact replicas of their pre-Columbian ancestors from Palanqué or Chichen Itza. For several millenia the Chinese have numbered hundreds of millions. The conditions of their physical and social environment have favored intensive selection. To what result? None. They simply remain Chinese. Within each population, men differ by their genotype, and yet the species *Homo sapiens* has not modified its plan or structure of functions. To the common base are added a variety of diversifying and personifying ornaments, totally lacking evolutionary value.

Of greater importance are the variations that some animals have undergone when confined to islands (such as Drepanididae (passeriforms) in the Hawaiian Islands; Geospizidae in the Galapagos Islands, Vangidae in Madagascar, Polynesian pigeons) or confined to lakes (such as the sponges, the planaria (90 species), and the gammarid amphipods (239 species) of Lake Baikal; the fishes of the genus *Haplochromis* of Lake Malawi (more than 100 species). The original stock has evolved into new species and subspecies (some of which deserve classification as new genera) and adopt a specific ecological niche. In isolated biotopes, *genotypic modulation* is greater than in populations living in wide-open spaces; it is much greater than that exhibited by the birth moths in industrial areas but remains within the framework of "microevolution"; it merely adjusts the genotype to the ecological niche. Darwinians say that, because the selection pressure is less intense in these isolated spaces, species poorly equipped for competition can survive and evolve. In particular, predation is rare or nonexistent. The apterous and unprolific pigeons and rails which lived in the islands of the Indian Ocean before man's arrival could not survive once hunted by man, by his domesticated animals (dogs, cats), and by his commensals (rats).

What we have said about confined environments is not presented as gospel truth and must not be generalized. When one considers the fauna of Lake Baikal in its entirety, one notes that out of 842 species living in the offshore waters 708 are endemic, whereas the coastal waters harbor only a common fauna comprising 400 species (excluding the Coleoptera, Hemiptera, Collembola, etc.). No fish is unique to Lake Baikal except in the family Cottidae, represented by 23 endemic and two ubiquitous species living in the coastal waters. Out of the 12 species of lamellibranchs, only three are endemic, and they all live in the deep offshore waters. The inability of some genera to diversify remains unexplained. Lake Baikal, because of its huge size (31,500 km^2), is more of an inland sea than a lake and therefore cannot be considered as a small environment where competition between species is slight.

In confined environments, it may be true that the selection pressure between species is low, but we cannot say with certainty that this is sure between the members of the same species. Lack of space and food are two factors which slow down demographic growth and also intensify the struggle for survival. The situation, which at first glance seems simple, appears to be more complex when studied in depth.

Natural selection is more intense in open environments, because it must offer more efficient resistance to the variations of the populations which live in them, and for this reason it favors a single genotype only. It plays a conservative rather than an innovating role. The mutations which diverge from the wild type or from the privileged genotype are swept away when the environment changes; hence the stability of the species.

Panchronic species, which like other species are subject to the assaults of mutations, remain unchanged. Their variants are eliminated except possibly for neutral mutants. In any event, their stability is an observed fact and not a theoretical concept.

Bacteria, the study of which has formed a great part of the foundation of genetics and molecular biology, are the organisms which, because of their huge numbers, produce the most mutants. This is why they gave rise to an infinite variety of species, called strains, which can be revealed by breeding or tests. Like *Erophila verna*, bacteria, despite their great production of intraspecific varieties, exhibit a great fidelity to their species. The bacillus *Escherichia coli*, whose mutants have been studied very carefully, is the best example. The reader will agree that it is surprising, to say the least, to want to prove evolution and to discover its mechanisms and then to choose as a material for this study a being which practically stabilized a billion years ago!

What is the use of their unceasing mutations, if they do not change? In sum, the mutations of bacteria and viruses are merely hereditary fluctuations around a median position; a swing to the right, a swing to the left, but no final evolutionary effect.

Cockroaches, which are one of the most venerable living relict groups, have remained more or less unchanged since the Permian, yet they have undergone as many mutations as *Drosophilia*, a Tertiary insect.

It is important to note that *relict species mutate as much as others do, but do not evolve, not even when they live in conditions favorable to change* (diversity of environments, cosmopolitianism, large populations).

How does the Darwinian mutational interpretation of evolution account for the fact that the species that have been the most stable—some of them for the last hundreds of millions of years—have mutated as much as the others do? Once one has noticed microvariations (on the one hand) and specific stability (on the other), it seems very difficult to conclude that the former (microvariation) comes into play in the evolutionary process.

Some contemporary biologists, as soon as they observe a mutation, talk about evolution. They are implicitly supporting the following syllogism: mutations are the only evolutionary variations, all living beings undergo mutations, therefore all living beings evolve.

This logical scheme is, however, unacceptable: first, because its major premise is neither obvious nor general; second, because its conclusion does not agree with the facts. No matter how numerous they may be, mutations do not produce any kind of evolution.

We add that it would be all too easy to object that mutations have no evolutionary effect because they are eliminated by natural selection. Lethal mutations (the worst kind) are effectively eliminated, but others persist as alleles. The human species provides a great many examples of this, e.g., the color of the eyes, the shape of the auricle, dermatoglyphics, the color and texture of the hair, the pigmentation of the skin. Mutants are present within every population, from bacteria to man. There can be no doubt about it. But for the evolutionist, the essential lies elsewhere: in the fact that mutations do not coincide with evolution.

EVOLUTION IS THE REAL HISTORY OF LIFE

In all respects, evolution is a long story. Spontaneous generation occurred once and only once; life cannot be reinvented, it is transmitted, it "is" continuity. Our cells are the daughters (to the nth generation, but daughters nevertheless) of the first animal which appeared on the surface of the earth some 800 million years ago; this animal was itself partly reproducing the substance out of which the first living being, floating in the salt waters of the primeval ocean, was made.

The study of the groups of animals or plants for which we have fossil evidence has revealed that, in their case, evolution is not the continuous unfolding of a simple phenomenon occurring at a regular speed and repeating itself in a regular sequence. It is a history, that is to say a maze of facts, of phenomena, pertaining to a group of objects whose nature, arrangement, and order change with time, following certain irreversible rules or laws. All objects, all matter, obey the laws of physics and chemistry. Any body placed in the vicinity of our planet is submitted to terrestrial attraction and, depending on its speed, will fall vertically to the earth. The laws of physics and chemistry are universal.

Yet, in the course of time celestial bodies undergo irreversible variations: a star cannot return to its original state; the conditions prevailing in the earth's mass and on its surface shortly after its genesis will never be found again. The macrocosm has its history which unfolds according to the laws of matter, that is to say the laws of universal determinism. The theory of the expansion of the universe, presently considered as the most plausible cosmogeny, is a fairly good explanation of the evolution of the worlds.

Geology is the history of the earth; cosmology is that of the macrocosm. On the other hand, physics and chemistry are not historical sciences; they are sciences of the matter whose laws are eternal and unchanging.

Indeed, the evolution of the macrocosm and that of the biocosm are based upon fundamentally different mechanisms and their respective laws are apparently unrelated. But do they not both reflect the instability of all that exists? Today's universe is not what it was yesterday, nor what it will be tomorrow. Without indulging in Spencer's evolutionism, which is outdated in its form as well as in its imprecise and too-simple principles, one must consider the facts and take them into account when making any scientific or philosophical interpretation of the living being.

The historicity of biological evolution is proved by the present complete interruption of all forms of spontaneous generation of living beings from inert materials. Creation from nitrogenous and other organic compounds dispersed or aggregated to a varying degree in sea water cannot be repeated. Spontaneous generation, which satisfies our logic, was a historical phenomenon in the highest sense of the word; although impossible today, it did, however, occur on earth in its early days or on another planet outside of our solar system, once only, and that was enough to launch life in the cosmos.

Although the macrocosm develops according to the laws of physics and chemistry, the living world adds to its history by obeying the same laws; it also follows its own laws which we know only partially.

CONCLUSION

Our logic, with its many hypotheses, attributes the interruption of biogenesis to changes in the physicochemical conditions prevailing on earth, around the earth, under the earth's surface, and in the seas, which prevent the synthesis of prebiotic materials. Once the proteins floating in the ocean waters had been consumed by the first living be-ings, the recurrence of any new biogenesis became impossible.

This situation required that their immediate successors possess the *ability* to reproduce on their own, as well as a *capacity* for chemosynthesis. Their perenniality could not have been maintained without these two conditions.

Here we are caught in the snare of science fiction and dream. Let's return to reality. What we are fairly sure of is that biogenesis *did not repeat itself.*

Unique phenomena, such as the genesis of the phyla, of the classes, of the orders, stand out in the past of living beings. Evolution proceeds from one historical fact to another.

The difficult task of the biologist is to discover the determinism of each of these historical facts. The causes which blocked the genesis of new types of organization are unknown. Were they changes of the terrestrial environment? Perhaps. Were they changes in the internal cell structure? Could this really be so? It is absurd to attempt to find an explanation for such complex phenomena in the absence of concrete data and experi-ments. Knowing of their existence is in itself an achievement. It will be possible to interpret these phenomena only when a better knowledge of genealogy and of the fundamental mechanisms of living beings has been acquired.

History appears as a series of events related to one another by so many causes that one can never be sure of knowing them all. It assumes an intricate structure, to which we have not yet been able to apply the techniques of cybernetics, because of the great number of parameters and their capacity.

The human historical phenomenon differs from the evolutionary his-torical phenomenon because man does not passively submit himself to the action of his physical, social, and economic environment. He acts in a forceful way and changes the course of events. This greatly increases the inextricable complexity of the determinism of historical phe-nomena.

As for evolution, it continues its course in a direction which neither Alexander nor Napoleon could deviate. Historical irreversibility is due to (1) the low probability of reuniting the same objects and of submitting them to the same physical and chemical conditions; and (2) the vari-ability of the causes and of their effects, which in turn become causes and thus change the nature and the order of the whole. For history to be reenacted, the conditions prevailing in the external and internal envi-ronments at its origin would have to be reproduced. This is theoretically possible, but in fact impossible.

From a historical standpoint, all evolutionary phenomena appear irreversible. Among these, the geneses of the major structural plans can never recur.

Can one say that the living world goes through a cycle, which is immense because of the great number of its participants and because of its length which to our eyes seems infinite? Perhaps. But this cycle would be different from that of a tapeworm or a trematode, which consists of compulsory and precisely defined stages and which, because of the reproductive function, is identically perpetuated. Evolution never starts all over again: the cycle of a parasite is not a historical phenomenon, but evolution is.

IV

Evolution and Chance

Ever since man has tried to understand the origin and nature of the universe, philosophers and scientists, whether avowed disciples of Democritus or not, have maintained and still do, that uncreated and eternal matter gave rise, through the incessant movement of particles, to the stars and all things, and that this happened by chance. Both life and the sidereal order sprang out of "chance": a minute bacterium and the greatest human genius are the sons of chance. Creation did not obey any laws, it was the realm of the "undetermined." No power, no higher principle nor intention was, or is, directing the creation of the living world, its changes and even less its possible advances.

Everything is matter and chance: no spirit or intelligence imposes this order. Yet, and this is a surprising paradox, the laws governing natural phenomena, revealed by physics and chemistry, originate from chaos. We wonder how Democritus and his contemporary disciples can reconcile chance and determinism. Isn't there an antinomy between those two concepts?

THE LOTTERY OF LIFE

What is the nature of chance that such great power is attributed to it? A single word often—if not always—conceals several different concepts. This is the case here. For the time being, putting aside philosophical speculations, we shall adopt the extensive definition appearing in the Littré dictionary: "Chance is incompatible with order and intent; it is a set of events independent of any cause or effect." This definition accurately applies to those biological phenomena considered by theoreticians to be random and not subject to a given law.

92

Exceptional, unforeseeable, or even inexplicable phenomena would hence be fortuitous. These very vague adjectives too often have a merely circumstantial meaning. A given phenomenon, today considered random, may tomorrow be considered determined because its causes will have been unraveled by thorough and specific study.

Biologists, whose task is not to seek moral causes or intentions, must first of all make sure that so-called random facts really are random facts; they must constantly keep in mind Poincaré's (1912b, p. 65) famous phrase: "Chance is only the measure of our ignorance."

It is a truism to state that biological phenomena are subject to universal determinism. How could it be otherwise? Can't they be ultimately reduced to physicochemical processes? We are not being ambiguous here, since we are certain that they can. The fundamental properties of living beings derive from chemical substances which range from ions to complex molecules; their architecture has no equivalent in inert matter. Thus, chance or randomness, as it occurs in living beings, only concerns the nature, the physical condition of materials and their mutual arrangement, with varying dimensional sizes and levels of complexity within a given structural plan.

Is it true that life was a fortuitous phenomenon most unlikely to take place? Biochemical and biophysical studies on the processes through which prebiotic substances were formed have led to the realization that this opinion is too extreme. Oparin (1967, p. 19), an authority on problems of biogenesis and an uncompromising materialist, states: "A remarkable peculiarity of living systems is that the whole network of metabolic reactions is not only strictly coordinated but also directed towards perpetual self-preservation and self-reproduction of the organism under a given set of conditions in the external environment. This highly organized orientation is characteristic of life and could not have been due to chance." Without being explicit, Oparin maintains that natural selection, operating as an antichance device, was the "ordering" agent.

Bogdanski, in various publications (1971, 1972), analyzes the vital phenomena in the light of physical, chemical, and biological data. He notes that these phenomena, far from being random, are highly determined and occur "following very complex laws which cannot be observed by a superficial study" (Bogdanski, 1972). Signs of life are grouped within a narrow band of physicochemical phenomena, outside of which there is no life. "As a rule, we can assert that the laws of "biorüge" (life in action) are based on a system of immanent prohibitions on each dimensional level. This excludes the probability of random events at the level of the phenotype." Such is the opinion of a biophysicist, not influenced by philosophical considerations.

But then the question to be answered is where does randomness spring from and when does it affect life? It has often been said that the fate of every multicellular and sexual organism is sealed by two successive lottery draws. The first draw occurs during the formation of the ovum or of the spermatozoon when the number of chromosomes is halved (since there is only one replication) and when the $2n$ chromosomes of the specific pool are distributed at random, but in equal numbers (n), between the two secondary spermatocytes, or between the secondary oocyte and the first polar body; this phenomenon is called meiosis or meiotic reduction.*

In the cell, chromosomes appear in pairs: one is paternal and one is maternal; they may contain alternative forms of a particular gene, called alleles. If the genotype is heterozygous, which is usually the case, this lottery, in the gametes, recombines the chromosomes in a manner different from that in the mother cells. Since all chromosomes are subject to the same intrinsic as well as extrinsic conditions, statistics can be applied to determine their distribution when they split in two and move toward the opposite poles of the achromatic figure of spermatocyte I or oocyte I.

The second lottery of ontogenesis is drawn during *fertilization*. The ova and the spermatozoa of heterozygous organisms do not contain exactly the same alleles. Their "half-genome" therefore varies according to the results of the first lottery. The fusion of the two gametes brings together two different half-genomes. Whatever the number of pairs of alleles —one or n—the probability of their combination follows certain mathematical laws which were discovered, more than a century ago, by Gregor Mendel, an Austrian monk. These precise laws usually work, if nothing in the external environment or in the reproducing cells produces physiological differences between gametes of the same sex. Some well-known causes sometimes upset the law by favoring certain gametes at the expense of others (selective fertilization); however, as a rule, the fusion of the gametes in a given species follows Mendel's statistical laws.

These two lotteries of sexual reproduction thoroughly stir up and mix

---

*One must keep in mind that in every diploid cell, or $2n$ chromosomes, n are maternal and n are paternal. The n chromosomes differ from one another: each one has its specific aspect and its own gene load. Thus, in the cell, "A" chromosome is present in duplicate (one paternal, one maternal). These "A" chromosomes are called homologous and make up a pair. If they have an identical gene composition, they are called *Homozygotes* and if they differ by one or more genes they are called *Heterozygotes*. The same applies to all other chromosomes except for a number of sex chromosomes which are nonhomologous. In the human species, the male chromosome sets will be $2n$ normal chromosomes, or autosomes, + XY sex chromosomes (= heterochromosomes); the female chromosome set will be $2n$ + XX. Other combinations can be found in animals and plants.

chromosomes and thus give rise to entirely new gene arrangements (if the sex cells are heterozygous), *but they do not produce any new character.** They provide sexual reproduction with the capacity to differentiate individuals descended from the same parents both chemically (namely, "personalization" of the proteins) and structurally.

In another book,† I have shown that statistical probability applies to events which occur by using elements which are qualitatively different while still obeying determinism. Mendel's laws are a very good example of this. But I wonder whether random is the right word to qualify these events, since the probability of occurrence of a given combination can be determined. Statistical laws are similar to any other laws; they enable one to forecast the outcome of events when they are not considered singly but in groups or series about which everything is known, since their constituents are finite in number. The outcome of such events is never unforeseeable.

True random events are quite different: they occur when the paths of two series of independent events having their own specific determinism meet and intersect (hence our use of the adjective "secant"), which they normally never do. This meeting produces a new event in which determinism is indeed present, but the meeting of the series depends on circumstances which are so varied that it is practically unpredictable. If the analysis of the phenomenon was continued far enough, its multiple and subtle causes would most likely be discovered. An example of this type of random event is the loosening of a roof tile by the wind, its falling off and hitting a passerby on the head. In biology, the equivalent is the meeting of a chemical or physical mutagen with the cell, this agent normally having no relationship with it.‡ DNA replication is thus upset; after the "attack," it will not produce a correct facsimile, but a faulty copy.

The secant random event is an accident, the statistical random event is not, but is subject to statistical, usually simple, laws.

The first type of random event differs from the second by its nature (two or three series of intersecting independent events) and by the considerable number of factors intervening in these series.

---

*The reciprocal exchange of segments between two chromosomes (crossing over) contributes to a certain extent in producing new gene combinations (see Strickberger, 1968).
†Grassé, *Toi, ce petit dieu,* p. 23 and following.
‡The mechanism of mutagenesis is still not well known. For our discussion, it is not relevant to know whether the ionizing or chemical agent scores a direct hit on DNA—as an arrow hits its target—or whether, on the contrary, it operates indirectly. All we need to know is that after the intervention of the mutagen, DNA no longer replicates quite identically.

CHANCE AND ITS INSTRUMENT, MUTATION, OR THE ERRORS OF THE GENETIC CODE

A typing error made in copying a text can be compared to biological mutation. According to advocates of Democritus, the DNA copying error would have unlimited fertility. From it would arise all novelty. Man would thus be the result of millions of errors which started a billion and a half years ago when the DNA of some schizophytes no longer reproduced identically and thus, by mere chance, produced the structure called the cell.

The "deviations" of the cell are unpredictable. This is true of the appearance and the effects of spontaneous mutations,* as well as mutations induced chemically (e.g., by mustard gas, formol, nitrous acid) or physically (ionizing agents: X-rays, radioactive matter, neutrons, ultraviolet rays) in plants or animals.

As a mutation alters the genetic code, it becomes automatically inscribed in the gene pool of the individual. If the original DNA is considered the "normal" model of the species, mutated DNA is thus abnormal. However, this is only an anthropomorphic appreciation of facts.

Strangely enough, although mutations are not predictable, their rate of appearance, which is almost constant, is predictable. A given dose of X-rays or any other mutagen, applied to a group of animals or plants, induces an increase in the number of mutants which remains more or less the same from one experiment to another. If the target theory (the mutagen scoring a direct hit on the DNA strand, as an arrow hits its target) is correct, the probability of mutation is of the same nature as in the game of heads and tails, or as that of a pin falling into a crack in a wood floor.

Moreover, we are sure that the same mutation can reappear several times.† This supports the idea that a gene can only change within certain limits. In other words, it cannot become any sort of gene. It is almost certain that the number of possible variations is limited. This conclusion

*Spontaneous mutations have been attributed to natural radioactivity. This is only one of many hypotheses; it can be granted some likelihood, but its relevance has not been proved.

†For example, in man, achondroplastic dwarfism caused by mutations occurs once out of 12,000 births (study carried out in Copenhagen on 94,075 children). The rate of hemophilic mutation could be 1 in 33,000 gametes. Congenital hip displacement, achromatopsia, albinism, etc., are all mutations which appear sporadically in human populations. Doctors, veterinarians, and agronomists observe these facts daily, and eugenicists take them into serious consideration.

has important theoretical implications. The number of mutations computed by geneticists is extremely high; however, the types of mutants are very much fewer in number. The source from which arises the evolutionary flow is less important than suggested by Darwinians. The "infinite creative potential" of DNA is surely not so great as has been claimed.

Mutations have a very limited "constructive capacity"; this is why the formation of hair by mutation of reptilian scales seems to be a phenomenon of infinitesimal probability; the formation of mammae by mutation of reptilian integumentary glands is hardly more likely (integuments of reptiles show very few integumentary glands; Gabe and Saint-Girons, 1967), etc.

The presence, in a given population, of individuals preadapted to new environments or to exceptional circumstances results from a sequence of lucky random events: (1) mutation; (2) full satisfaction of a preexisting need by the mutation; (3) dominance of the mutation or the homozygous state of the mutant; (4) timely appearance of the mutant. The probability of preadaptation decreases as the number of conditions required for success (i.e., parameters and circumstances) increases.

Gene variations are not the only ones which modify the cell structure. Mutations due to chromosomal rearrangement (translocations, inversions of segments, exchange of segments between chromosomes, breakage and reunion of chromosomes) are hereditary and play a part in the differentiation and separation of races, of subspecies, and even of species. They produce new genetic arrangements without alteration of the genes.

To insist that DNA copy errors are the only hereditary variations producing new characteristics is tantamount to admitting that biological evolution was not subject to determinism and occurred totally at random.

BIOLOGICAL ORDER AND ANTICHANCE

Mutations, in time, occur incoherently. They are not complementary to one another, nor are they cumulative in successive generations toward a given direction. They modify what preexists, but they do so in disorder, no matter how.

Derivation obviously does not demand that variations be coherent and follow preferential paths.Evolution, however, followed courses to which it faithfully adhered over very long periods of time. Although everything is not as it should be, the living world is not at all chaotic and life results from a very well-defined order.

As soon as some disorder, even slight, appears in an organized being, sickness, then death follow. There is no possible compromise between the phenomena of life and anarchy. Disorder does not alter or endanger the hierarchy of organs or functions in any living being. Apicobasal gradients, morphogenetic fields, and "cephalization" bear witness to the interdependence and the hierarchy of the various parts of the organism. Any ontogeny of a multicellular animal or plant implies that a hierarchical organization is set up in the definitive form of the species; this organization is already potentially inscribed in the fertilized ovum which contains the entirety of the specific properties.

Individual or racial genetic variations are quite different: They appear in a disorderly manner and do not show any preferential tendency. The presumed correlation between black skin, fuzzy hair, and everted lips of the Negro race is of no biological value; nor is the combination of epicanthic eyefold, yellowish complexion, and mongoloid spot of Mongoloids.

Confronted with the evidence of order, of adaptation of forms to functions, and of the harmony between ways of life and environment, theoreticians of "random evolution" backpedal and reduce the importance they give to all-powerful chance. Randomness *must fall back into an order*, i.e., or no longer be randomness. For this, Darwinians have recourse to natural selection which, according to them, is the most efficient antichance device and the "organizing" agent of the vital structures and functions. Thanks to natural selection, chaos is replaced by a system so harmonious in its parts and in its whole that they indulge in thinking that it is due to causes still beyond our understanding.

In order to measure the part played by chance in the genesis of life as well as in its evolution, it is necessary to take a closer look at what mutations really are and what their role is in the genesis and development of new types. We know that they are spontaneous, unpredictable, hereditary and unspecific in any given function or structure, and most often lethal. Selection acts on mutants following Malthusian parameters (Waddington, 1965) which some authors designate by their result, *differential mortality*; this eliminates mutants bearing characters unfavorable to the species.

Statistical probability applied to demography and to natural selection has permitted not only a better analysis of phenomena, but also an evaluation of their long-term effects. However, the mathematical theory of natural selection must not delude us. It is only worth what the principles upon which it is based are worth. Our imperfect knowledge of the factors governing the survival or the death of the members of a given population greatly reduces the precision and the value of theoretical forecasts. For example, the final outcome of the competition between two kinds of mutants has been determined statistically; the mutants used in that experiment belonged to a population of Drosophila living in a confined space under specified conditions, and it was assumed that mating was random (panmixia). In fact, living beings rarely mate at random; promiscuity in higher organisms is exceptional, and it is even doubtful that it exists. Drosophila also follow this rule and choose their mate according to sensory data and inborn responses. Therefore, theory and calculations had to be reconceived. The discovery of any new factor makes it necessary to reconsider evaluations and to modify the statistical laws supporting them.

These qualifications do not relate to the fundamental problem, which is to determine the "actual" (and not the presumed) role that selection held, and still holds, in evolution. The following chapter is devoted to the analysis of the conditions under which the problem arises and to its solutions.

As regards the antichance function of selection, one must keep in mind that the selection coefficient of a given character varies from one environment to the next (in the broadest sense of the term environment). Theoretically, if the environment is stable, the population should not change, since it is in equilibrium. Disturbances only occur when there is a change in the external conditions in which the given population is placed; this change must be powerful enough to throw the population out of balance, otherwise it has no effect.

Selection adjusts the genotype of the individual to its environment. The mutants of a population provide it with some chances of adapting more precisely to the circumstances. As mentioned above, mutations and natural selection act as a stabilizing mechanism. But whereas in homeostasis the return to equilibrium results from a reaction of the organism, in this case, the genotypic stabilization occurs without involving any intervention of the living being: it undergoes mutation, then selection.

Thus, although submitted to the randomness of mutations, the living being, owing to an antichance factor having a precise orientation, preserves the right genotype or changes it in order to suit its needs. From one genotypic state to the next, from one random mutation to the next, evolution follows its course (at no time does the living being intervene), thanks to the filtering role given by selection to "usefulness" and "adequacy."

RANDOM PREADAPTATION

The major achievements of molecular biology lie in the fields of genetics and cellular metabolism. Genetics is the science of the *heritable*, of the *transmittable* with respect to similarity, and as a by-product only, the science of the unstable, of the variable. *A strictly stable state would be incompatible with the survival of living beings*, of populations or species living under different climates (torrid, freezing, or temperate).

We know that a single gene can be expressed under several forms or alleles which are the result of its mutations. The homozygous condition of a population (or rather something which approaches it since it is practically impossible to maintain that a living being is homozygous in all its genes) is extremely rare and unfavorable to the species. Logically, the more heterozygous a population is, the more chances it has, thanks to its store of diversified genes, of possessing the character and the property that enable it to adapt to environmental changes, to new circumstances. For the species, the heterozygous condition is thus an "insurance" against future disasters. Despite its blindness, chance becomes the tutelary god of the population. For example, because of chance, mutants capable of resisting sulfanilamides or antibiotics can be found in cultures of gonococci; in populations of pneumococci, the resistant mutants are those producing an enzyme, penicillinase, which inactivates penicillin, etc. This victory of the species over the drug was made possible by the appearance, prior to exposure to antibiotics, of random mutants producing resistant strains. *Preadaptation* is one of the names given to this random insurance of the species; *the randomness of mutations offsets the randomness of changes in environment and in circumstances and prevents their deleterious effects.*

The entirety of the alleles of a given population constitutes its stabilizing system; it comes into play when the harmony between the organism and the environment is, or is about to be, destroyed. Therefore, it is a balancing system and not an innovating one. Although in most cases the mutant is less fit than the invariant, this situation can sometimes be reversed and the mutation, once deleterious, becomes beneficial.

However, one must not forget that genetic variation, in single or double dose, is most often lethal to the animal or plant in which it occurs. The long list of family illnesses, which in man and animals originate from mutations pathogenic enough to cause death, is very instructive in this respect.

Geneticists, from the sequence of four nucleotides, which are the monomeric units of DNA molecules, worked out the number of all possible alleles found in the chromosomal DNA of a given species. The numbers obtained were extremely high but uncertain. Up to now, the number of mutations observed in the thoroughly studied animals and plants (*Escherichia coli, Neurospora, Saccharomyces,* corn, tobacco, cotton, rat, mouse, rabbits, household pets, man) has not exceeded a few thousand for a single species, but it is likely that their number is in fact much higher (see below).

Their computations did not give due consideration to the fact that not all genes mutate at the same frequency. "Archaic" genes, which determine the proteins and fundamental functions (Krebs cycle, for example), apparently mutate less often than more recently acquired genes. There is a possibility that some of the larger changes are incompatible with life and thus lead to the very early death of the embryo; the mutation then remains unnoticed.

Genetics has revealed the rates of spontaneous mutations in some species. In *Drosophila,* for example, in the second chromosome the rate of lethal mutations reaches 0.5%, in the third chromosome a little less than 0.5%; it is very much lower in the fourth small chromosome. In the first chromosome or sexual chromosome, the rate of sexually related recessive mutations is 0.1%.

Highly lethal mutants represent 1% of the population. If one adds this 1% to the numerous mutations with slightly lethal effects and to the rare mutations beneficial to the species, the total reached is 5%, i.e., one mutation in every 20 gametes. This extremely high ratio would be higher still if mutations affecting polygenes were taken into account; it would then involve 35% of the gametes (Mukai, 1964). If *neutral mutations* (which modify the structure of proteins to such a small extent that they are neither beneficial nor deleterious to the mutant) are taken into account, the ratio is even higher. One can admit that the number of preadapted individuals is proportional to that of the mutants; little can be learned from quantitative data in respect to the "quality" of the mutant—this will be revealed only when the individual is confronted with reality.

No matter how large the number of mutants, no one can say whether the one capable of overcoming the disturbances caused by the environmental and ethological changes and by aggressions of all kinds is present among them. The three billion living people and the billions that have passed away never underwent the mutation which would have made them resistant to plasmodia, which cause malaria; yet, this would have been so useful for the human species! The same applies to all the other parasitic and infectious diseases. Genotypic regulation is part of the Darwinian antichance scheme. Its scope is limited since "useful" mutations are rare or extremely rare.

CHANCE AND EVOLUTIONARY TRENDS

A lineage is a series of species descending one from another and heading toward a given type. If oriented lines did not exist, the living world would be chaos. There would be no need for evolutionists since what is incoherent does not fall into the sphere of science.

It is worth mentioning that the concept of evolution is historically related to that of orientation. Lamarck (1809) prolcaimed that the animal kingdom progresses from the simple to the complex (see Lamarck, 1873 edition, Vol. II, p. 424, his table showing the phylogeny of the animal kingdom). Darwin (1859) considered natural selection to be the agent of biological progress for the greatest benefit of the individual and the species.

When reference is made to fossils, evolution always appears as a continuous approach toward a given form. It occurs in a single line, by the adding of successive, complementary variations going in the same direction for millions and millions of years. For instance, the genesis of mammals from reptiles lasted more than 50 million years.

Paleontological evidence does not support the evolutionary randomness doctrine: evolution follows the same general direction and keeps to it as long as the line concerned has not fully attained a certain form, its idiomorphon. Yet, some paleontologists (e.g., G. G. Simpson) are unconditional advocates of the randomness doctrine, and oppose the concept of oriented evolution which they wrongly mistake for orthogenesis.

We wish to stress that *the existence of oriented lines is a fact, and not a theoretical view; a line can only be identified and exists solely because it embodies a given trend appearing in individuals which derive from one another and succeed one another in time.* Species A becomes B, then C; with each new form, certain features of A become increasingly marked, and so on. In actual fact, things are more complex.

As established in Chapter II, evolution is highly branching; however, it is not disorderly. Each branch follows its own trend while respecting the general evolutionary theme of the entire order (Primates, Proboscidia), superfamily (Equoidea), etc. The lesson to be learned from paleontology is that no phylogenesis has taken place in a disorderly manner.*

---

*It must be understood that to discard oriented evolution means discarding simultaneously progressive, regressive, or coordinated evolution. How could the wings of insects and birds have formed without going through initial imperfect forms? The least we can say is that the idea that they could have appeared at random and have been functional from the outset is preposterous.

Lethal mutations, by far the most frequent, are excluded from any evolutionary process.† Recessive mutations, those not expressed in the phenotype, do not undergo selection which acts only upon an actual condition and not on a potential one. Some have been tempted to think that the recessive gene, when combined with other genes, can become dominant and selective, but the mechanism and frequency of this varying dominance is not well known.

The opportune appearance of mutations permitting animals and plants to meet their needs seems hard to believe. Yet the Darwinian theory is even more demanding: A single plant, a single animal would require thousands and thousands of lucky, appropriate events. Thus, miracles would become the rule: events with an infinitesimal probability could not fail to occur. Much as in *The Swiss Family Robinson*, which I used to read in my childhood, rescue would always occur at the right moment, and this would have had to have happened throughout the ages. One could admit that one bacterium out of billions and billions can be the "lucky preadapted" one, but the number of reptiles evolving into mammals or of primates evolving into men, did not exceed a few tens of thousands and often fewer; the chances of the appearance of "useful" mutations therefore decrease in the same ratio and become almost nonexistent.

Darwinians answer that we consider only the successful mutants and disregard the failures, since they disappear. This is only partially true. Extinct lines are not in the majority, which means that successes (agreement between need and mutation) have prevailed and that, consequently, the presence of useful preadapted individuals was the rule. One can see how far this might lead.

Darwinians rarely and discretely mention the possibility of finding in the allele pool of an unspecific population the gene capable of satisfying a given need of the species. They know, of course, that this possibility is

---

Moreover, from what we know of the mechanism of mutations, invention in biology has never been the product of a genetic variation; it can occur only through the combination of several changes. Thanks to the coordination of parts, the whole is fully functional. It materializes a structural plan, the origin of which is totally unknown; natural selection could not have conceived it, and even less constructed it since adequate materials were lacking.

Simpson's opposition to oriented evolution leads ultimately to the denial of evolution: in the living world, everything would be subject to randomness or change and would occur in just any fashion, and at any time. Simpson offers chaos whereas in fact the living world has evolved and perpetuates itself in an orderly manner.

†Sickle cell anemia, which is caused by abnormal hemoglobin, is most likely due to a mutation; its pathological character notwithstanding, it protects man against malaria.

very low, but this does not bother them since, they say, millions of present species managed to get through in spite of their narrow chances of success. In any event, this extremely low chance of success certainly contrasts with the constant success of the species and should puzzle the least curious among us.

What gambler would be crazy enough to play roulette with random evolution? The probability of dust carried by the wind reproducing Dürer's "Melancholia" is less infinitesimal than the probability of copy errors in the DNA molecule leading to the formation of the eye; besides, these errors had *no relationship whatsoever* with the function that the eye would have to perform or was starting to perform. There is no law against daydreaming, but science must not indulge in it.

Thus, preadaptation, according to Darwinian theory, is equivalent to a potential preestablished harmony in certain animals and certain plants. Throughout the living world, mechanisms, organs, and enzymes, scattered in organisms, are at battle stations waiting for an event, random of course, to complete what is missing, to protect what must be protected and to neutralize what must be neutralized. Philosophers will draw from this picture the conclusions that suit them. Leibniz would probably find in it support for his conception of life.

CHANCE AND ORGANIC COMPLEXITY

Leaving fossils and "preadaptation" aside for the moment, let us now define the conditions necessary for mutations to contribute to the genesis of complex organs; these conditions are opportune occurrence, harmony with preexisting conditions, and coordination with other mutations.

Our study will concentrate on the eye, the genesis of which is a major challenge to evolutionists. Julian Huxley, at a conference held at the Palais de la Découverte in Paris in 1946, declared that the problem was solved. We consider this statement to be overly enthusiastic.

Charles Darwin, more cautious than his disciple, recognized the weaknesses of his theory, which are increasingly apparent today. We are not surprised, then, to read in a letter to his friend the botanist Asa Gray: "To this day the eye makes me shudder, but when I think of the fine known gradations, my reason tells me I ought to conquer my fear" (Darwin, 1800, p. 273, letter to Asa Gray, February 1860).

We fully understand Darwin's fears and wonder what they would have been, had he been confronted with the anatomical and cytological complexity that is revealed by modern biology; he would have been even more worried had he known that selection cannot create anything on its own. We know absolutely nothing about the evolution of the eye of the vertebrate, and embryology is of little help. The problem is to know whether random mutations could have given rise to an organ requiring, because of its complexity, a considerable number of data for its elaboration. The number of mutations must have been enormous for adequate ones to occur at a given point, by chance and to enable the organ to function.* We need not belabor the diversity of the transparent parts, on the relationships between the intraocular fluid (aqueous humor) and the venous system (Schlemm's canal), among others. The complexity of the retina, of the sheaths, etc., need not detain us either; all this is extremely well known, but we must say that no recent publication inspired by Darwinism even mentions it.

In 1860 Darwin considered only the eye, but today he would have to take into consideration all the cerebral connections of the organ. The retina is indirectly connected to the striated zone of the occipital lobe of the cerebral hemispheres: Specialized neurons correspond to each one of its parts—perhaps even to each one of its photoreceptor cells. The connection between the fibers of the optic nerve and the neurons of the occipital lobe in the geniculate body is absolutely perfect. The processes of the axons, the outgrowths of the dendrites, and the connections with corresponding elements are so precisely laid out in time and space that as a rule everything works perfectly.

In fact, the picture we have just sketched is even more complex; we did not consider the molecular structure which shows as many peculiarities of adaptation as the macrostructure (the subtleties of which were sometimes mistaken for imperfections; see Ivanoff, 1953), and we have neglected entirely the chemistry of a complex organ capable of multiple adjustments.

We took the eye as an example, but the ear would have been just as instructive. Is not the human brain, the organ capable of abstraction, an even better example? Even the architecture of the cortex with its 14 billion neurons is not known with any degree of precision. In mammals, *all sense organs evolved almost simultaneously.* If one considers the great number of simultaneous, timely mutations satisfying existing needs involved in their genesis, one can not fail to be confounded by so much harmony, so many lucky coincidences, due entirely to omnipotent chance.

*Darwin (1859) devotes four and a half pages of the "Origin of the Species" to the eye and its genesis, possibly thanks to innumerable mutants, to natural selection, and to time. But we note that he does not overcome any of the obstacles raised against his doctrine by "reality."

Selection must complete its work on successive generations and must find in them the materials it needs.* Moreover, successive generations reproduce preceding ones, otherwise they have no evolutionary value. We have already listed the lucky chances required for the slightest evolution to result from mutations (p. 94).

Anyone who endorses the random theory of evolution admits that the eye and the ear, to become what they are, have required thousands and thousands of lucky chances, synchronized with the needs of their construction. What probability is there of such wonderfully fortuitous success?†

Natural selection, if one admits that it is the builder of the living world, can only operate if it possesses the correct building materials needed for the construction of the organ at the right moment. What is the use of appropriate mutations if they appear too early or too late in the course of phylogenesis?

If the formation of the crystalline lens and of the retina had not been closely coordinated (the retina is the inducing agent of the anterior parts of the eye), the eye could not have formed. The necessary mutations could not have occurred independently. The influence of the organ extends to structures in its immediate vicinity; can one imagine an eye without eyelids or without lachrymal glands? Moreover, these accessories necessarily formed early in the course of evolution; the eye is indeed too fragile to be able to do without them. The chronology of phenomena in any ontogenesis is inflexible. The formation and the subsistence of the living being requires that successive transformations arise in an orderly manner and that its architecture be equally ordered. Randomness and chance have no place here.

Moreover, during phylogenetic organogenesis, natural selection must be capable of foresight. Isn't "choosing" its prime function? But the choice cannot take place without predicting the future role of the incipient organ. Without such prescience, the coordination of successive states is incomprehensible. Did Darwin take this into consideration?

Without its predictive powers, selection would not be able to favor an incipient organ which, at the time, had little or no usefulness. What sort of advantage could result from the starting of an eye, when the materials

---

*I have written this as if selection were an operating entity; my apologies to the reader. In doing so, I have borrowed from Darwinians their language and concepts.

†Salet (1972), through the use of mathematical methods, pertinently and severely criticizes the theory of random evolution. Although we do not share the biological and philosophical views of that author—who is not a biologist—we recommend that those who are interested in the role that chance could have played in the genesis of the living world read this book.

forming it were not yet transparent? Of what use was the development of the dentary and the accompanying regression of the proximal jaw bones in theriodont reptiles, the ancestors of mammals? An answer can always be invented, but all this merely adds another supposition to the mass of previous suppositions.

We repeatedly hear that chance is all-powerful. Statements are insufficient. Evidence must be produced. I do not consider the spontaneous appearance of resistance to an antibiotic in a nonresistant population of bacteria as evidence. Neither structures nor fundamental functions are involved here. This is so true that variations of this kind, although repeated millions of times, have left bacteria practically unchanged.

The living being, which must constantly maintain its functional unity, is a self-regulating system. Its intrinsic finality, perseverance in the act of being, is an essential property which cannot be concealed, even when given a name which is only a subterfuge.

Biochemists and physicians who are attempting to reproduce the accomplishments of biogenesis know that to *give life to a material system is to give it a purpose maintained durably by multiple self-regulations.*

On the border of science, however, a theoretical system gradually appeared which claimed that chance accounts for the genesis and the evolution of the biocosm. Its advocates have faith in chance; they are certain that it unfailingly provides the living being with all that it requires.

> It gives the nestling birds their food
> and all nature shares its bounty.*

Directed by all-powerful selection, chance becomes a sort of providence, which, under the cover of atheism, is not named but which is secretly worshipped. We believe that there is no reason for being forced to choose between "either randomness or the supernatural," a choice into which advocates of randomness in biology strive vainly to back their opponents. It is neither randomness nor supernatural power, but laws which govern living beings; to determine these laws is the aim and goal of science, which should here have the final say.

To insist, even with Olympian assurance, that life appeared quite by chance and evolved in this fashion, is an unfounded supposition which I believe to be wrong and not in accordance with the facts.

*Aux petits des oiseaux, il donne la pâture
Et sa bonté s'étend sur toute la nature.

V

Evolution and Natural Selection

Evolution as imagined by the Darwinian school, using mutants as its building blocks, cannot do without chance, which provides it with its material, but the selection to which evolution resorts in order to adjust mutations to necessity or simply utility becomes antichance. The keeper of order. Thus, in its most elaborate form the doctrine endeavors to reconcile opposing elements, chaos and order. It consequently links adaptation and selection tightly together, the former depending upon the latter. This conception affects both the way in which the usefulness of the characteristics of living things is to be judged and the genesis of the adaptation is interpreted. Therefore, any critique of the doctrine should relate to both of these notions; we shall now devote one chapter to each in turn.

THE NOTION OF SELECTION

The image of selection sifting through the variants in a given population, sorting out the fit from the unfit, was expressed long before the nineteenth century by a great many naturalists or philosophers. Aristotle, that universal precursor, even enunciated the principle of the struggle for life: "The animals are at war with one another whenever they live in the same places and take the same food. If food is scarce they fight, even when they belong to the same species." He even asks, in the "Physics" (Book II, Chapter 8), "whether such fighting may not have caused the extinction of forms insufficiently adapted to living conditions, and the conservation of those which are so adapted, whence the apparent *finality* we observe." But he at once rejects the idea, seeing that finality is in nature as the exception and not the rule; moreover, he believes the resources of nature are great enough to make it impossible for any one of its works to be destroyed. What is more, not all animals fight one another, some are friends" (Perrier, 1896, p. 16).

In other authors of antiquity we find allusions to competition among animals and the survival of the fittest. But nobody went on to deduce from this the general principle; the notion of evolution was lacking.

Darwin and Wallace (1858) were really the first to arrive at the idea of attributing to the struggle for existence a selective function in bringing about the evolution of living species. In their joint explanation of the mechanism of evolution, by far the most original idea is that survival is only for the individuals who are fittest to fight their fellows to stand up to their enemies of all kinds, to the rigors of the climate, to find food and resist infectious diseases, and so on. They found the inspiration for their idea, and arguments in favor of it, in the physiocrat Malthus' (1798) *Essay on the principle of population* (cf. Wallace, 1908), which drew attention to the existing imbalance between the population growth and the available food supply and to the causes of depopulation.

Natural selection remains the foundation of Darwinism, which postulates its *universality* and makes of it the agent responsible for the evolution of all living organisms. Only viable forms have survived (a truism), only those systems which perform the function required of them. Instead of the disorder of random mutations, selection creates order, equilibrium, even harmony.

We must now show the accuracy of this interpretation. To say of a given set of objects or living creatures that they are in equilibrium with their environment, or their components in relation to one another, does not mean that the equilibrium and overall state are the result of prior selection or choice. For instance, the present state of the macrocosm and its heavenly bodies has by no means come about by a process of choice, but by transition from one state to another, the change continuing until a certain stability is achieved, which in turn will be upset by internal or external causes.

If we say selection, we finalize the system. There can be no selection without a purpose. Whether it is willed by necessity or some other factor matters little; the causes vary, but are directed nonetheless. What interests us for the time being is to what extent the losses suffered by any animal or plant population participate in the evolutionary process.

SELECTION DURING ONTOGENY

Both animal and plant species suffer enormous losses, primarily affecting the reproductive systems and the embryo. An exceedingly high fertility rate notwithstanding, many populations remain numerically stable, every pair replaced in due course by another. This testifies to the high mortality rate among gametes, embryos, and the young in general.

In the case of animals among which the act of reproduction brings together hordes of individuals, destruction turns to massacre. The countless termites and ants, swarming out of their nests, are frantically attacked by predators. The only pairs to survive are those who manage to find shelter in the earth or behind a ridge of bark, quite irrespective of the specific qualities of the insect, and not in the slightest as a result of any kind of selection. The swarms of polychaete annelids emerging for mating purposes equally fall victim to innumerable and insatiable natural enemies. The wholesale destruction of eggs, spermatozoa, seeds, and larvae is not selective. Death does not choose its victims, but strikes blindly.

Deformed gametes, of which there are a great many, seldom actively influence reproduction because of their functional disability. But an ovum or spermatozoon may carry lethal genes and produce an embryo which develops normally until those genes come into play, in some cases at the latter end of ontogeny. Statistics show that in the human species 25% of abortions and stillbirths are due to a chromosome abnormality (as often as not a polysomy); of the other 75%, a high proportion quite likely results from the presence of lethal genes. During development of the embryo and in infancy, the elimination of the unfit, and of the pathological, is fully operative; it safeguards the genotype, but has no guiding influence in evolution.

SELECTION AND ACCIDENT

The massive losses caused by natural cataclysms that destroy huge areas are unselective, whether for animals or for plants. They devastate blindly and are random as to place and circumstance: tidal waves, floods, forest fires, bush fires respect no one and nothing.

The hurricane that uproots whole tracts of a forest has no truly selective action, but by felling trees it creates an environment favorable for some species and fatal to others. In the equatorial rain forest the huge scars left by tornados are empty spaces in which interspecific competition is intense and plant groups follow a cycle whose constant goal is simply the reversion to primary forest, its ecological climax.

The annual bush fires set by the natives of the African savanna during the dry season do spare a few species of plants called pyrophytes, whose thick bark, subterranean organs, and succulent tissues are capable of withstanding heat and flames. Thus, "preadaptation" (a term coined by

Lucien Cuénot, see Cuénot and Tétry, XX, pp. 413–424) favors certain species which, after the fire has passed by, become predominant. In the central African grassland, species of *Hygrophila* and *Lepidagathis* (of the Acanthaceae) are two such examples, low plants whose seed capsules are hard to open; before the fire the fruits stick together in lumps, allowing the flames to lick them, but they are so changed by the heat that at the first touch of dew they burst apart and scatter the seeds, thus enabling the plant to spread and flourish.

Of course the fire does affect the composition of the flora. But in the absence of specific research on the matter, all we can say is that within the same species it favors one genotype rather than another. *At any rate it does not call any novel species into being.* Selection of the preadapted individuals modifies the composition of the floral population but does not constitute true evolution, for the individuals remain unchanged. As soon as the periodic burnings end, the selective factor no longer operates; the pyrophytes are no longer privileged, and the population reverts to its prior condition before the fire started.

Let us not confuse creative evolution with variations in the composition of a population through circumstances. They are two distinct things, and any attempt to connect them is purely specious.

The examples of preadaptation that have been suggested so far (selection by frost of individuals able to withstand the cold in a plant population, by bush fires of pyrophytes, by variations in salinity of aquatic crustaceans, or by the wind of apterous insects, etc.), deal only with characteristics that do not affect organization and never set the species off on a new path.*

Thus, mass destruction strikes at random, without rhyme or reason. After the cataclysm, any preadapted individuals are the sole survivors, and the flora and fauna are altered in their composition, but there is little change in the individual's genetic status, and no large-scale variation occurs.

---

*Preadaptation does not always have the effects that are claimed for it. While species of wingless Deptera are more common by the sea and on islands than elsewhere, because they catch the wind less, there are wind-swept regions where there are only winged flies; this is the case of the huge "regs" of the Sahara, where there is nothing to stop the simoom and the sandstorms.

On the island of St. Paul in the middle of the South Pacific all Diptera normally have wings. Given the frequency among them of the apteran mutation, it is hard to see why this has not occurred among the island's populations. The answer is easy: When the wind blows harder than a given velocity the insect takes shelter, remaining close to the ground. Any entomologist who has gone butterfly hunting in the Mediterranean *garrigues* knows that when the mistral blows no insects fly.

SELECTIVE ELIMINATION

Accidental deaths are important to the growth, decline, or equilibrium of a population, but hardly affect evolution at all. After a random massacre there begins, in the surviving population, more or less fierce competition between larvae, juveniles, and adults, theoretically resulting in the survival of the fittest, or, in genetic terms, the genotype best adapted to the environment. Elimination is now, instead of random, selective.*

Survival of the fittest is the result of action that is essentially suppressive. If it operated fully, natural single-species populations would tend toward a unified genotype, and multispecies populations would tend toward monospecificity. But the not-so-good persists, the natural populations remain (genetically speaking) very highly heterogeneous.

So we need to find out how natural selection works. Until now more study has been devoted to experimental populations than to natural ones.

Although in the vast majority of cases the mutant that is generally "inferior" to the wild type is eliminated, the mutant persists and finds a niche in the population. This is what happened with an allele of *sepia* (eye color), observed by Teissier (1943) in laboratory *Drosophila*;† *sepia* gradually multiplied until it stabilized at or about a fixed frequency for some twenty generations. Demographic studies on experimental popu-

---

*A mathematical study of natural selection has tempted some mathematicians, the most famous being Volterra, and biologists such as Fisher (1930), Haldane (1924, 1924–1931, 1932), Teissier (1958), Sewall Wright (1969), and Kimura and Ohta (1971). Anyone seeking further information should consult these authors' publications; references to the principal ones will be found in the bibliography at the end of this book.

The Hardy–Weinberg law is merely mentioned as demonstrating the gap between theory and reality. It has been variously stated, and we quote it as follows: "In a large stable population in which matings are random (panmixia), selection is inoperative, and mutations do not occur, the frequencies of different genes and genotypes will remain constant in succeeding generations."

This law refers to an ideal state, not to any real conditions. Hence it is of little interest to the evolutionist, who has to consider concrete reality—what exists, not what is fictitious. Demographer-geneticists use it as a term of comparison between what they observe in breeding populations and the theoretical, stable states, permitting the introduction of new parameters and coefficients into the original equation, and the calculation of gene frequencies in the unstable state.

†In the caged populations in which the *Drosphilia* larvae develop, "the selective factor . . . is very fierce competition for food." In nature the competition is likely to be not so fierce, but fermenting fruit is not all that plentiful (except in the autumn, in vine-growing regions) and in all probability a great many adults perish without offspring. These randomly caused deaths have no selective value.

lations of *Drosophila* have revealed that selective values of genotypes depend upon the environmental conditions in which they live.

There are various ways of recording movements of living creatures. We present one that is particularly valuable for the evolutionist. It is taken from G. Teissier, the biologist who, in the author's view, has achieved the deepest insight into natural selection but whose Darwinist credentials are impeccable. He writes (Teissier, 1958):

> The frequency, in experimental populations of *Drosophila*, of certain mutants may stay practically constant for several tens of generations but if the experiment is continued long enough the frequency will normally undergo a succession of periods of stability, growth, and decline that may differ widely in level. These unexpected and swift variations, which determine the character of the evolution* of each of the populations studied, cannot wholly be explained either by unobserved changes in environmental conditions or by "genetic drift" caused by the mainly random character of the choice of gametes that will produce the new generations. It is permissible to think that they are due to unperceived changes in other systems of alleles than those under observation."

The environment being kept stable, we have to look for the causes of variations in the rate of mutations in the very nature of the individual members of the population. The explanation proposed by Teissier is likely to be the right one. Neutral and other unperceived mutations, which unceasingly differentiate parents and offspring and one individual from another, modify the selective values of genotypes, possibly by some sort of summation effect that accounts for their changing frequencies.

Taking now an evolutionist view of the matter, experiments and observations with experimental populations in a constant environment show the extent to which selection varies in its action, due to many different causes. Outside the laboratory, in the wild, the complexity of the environment increases considerably in proportions not easily quantified; both genotypes and selective factors increase in number.

It is difficult to tell what foothold a given characteristic may offer for selection. If the mutation endangers the life of the animal or plant, its effects on the population concerned are easily known. Of course, no measurement of the advantages or drawbacks of a given characteristic to its bearer makes sense unless it compares the respective numbers of offspring in which it is found or not found.

In the case of a single characteristic and in homogeneous populations, we can say that the differences in numbers do relate to the differential

*The term evolution is used here to mean history or cycle, and not in its transformist sense.

characteristic involved. In heterogeneous populations genic actions and interactions are so complex that it is hard to make such a statement.

Since species differ greatly in respect to the number of genes, any comparison to establish the selective value of a characteristic is practically meaningless. Demographic inequalities have too many causes for us to know which one is the more or less important.

The advantage or disadvantage resulting from a characteristic whose incidence is small is not easily assessed. For example, statistical studies on populations of the woodland snail (*Cepaea nemoralis*), carried out in France by Lamotte (1951, 1966) and in Britain by Cain and Sheppard (1950, 1952, 1954) and Cain and Currey (1963, 1968a,b) failed to reach the same conclusions. Lamotte considers the presence or absence, and relative size, of dark bands as nonselective. The English authors disagree with this conclusion and attribute the different occurrence of individuals with or without bands to selection. Despite the contribution made by Ford (1971), the debate has reached no definitive, reliable, or satisfactory conclusion. The opposing data are even more interesting because all the authors are orthodox Darwinians. It must be remembered that *Cepaea* shells found in Pleistocene deposits (about 1 million years old) already had dark and pink bands (Diver, 1929). This fact alone shows how unimportant adornments may be for the survival of the species: Let history be the judge, and its verdict is unmistakeable—survival or extinction.

Since evaluation of what is or is not advantageous is impossible in the case of fossil animal populations, whatever may be said about the selective value of a given characteristic is pure imagination. It is not because individuals with long spines become more numerous in a population of cidarid sea urchins that the characteristic "long spine" accounts for their predominance; this might be a very natural effect of growth continuing with age. Quite another characteristic (resistance to parasites, lower embryonic losses, etc.) may be a possible cause. Where the imagination is given free rein we must learn to control it.

Although the future of an experimental population remains unpredictable, despite all the high quality of the mathematical tools at the demographer's disposal and his mastery of the environmental parameters, this is much more the case in a natural population, where the prediction is virtually impossible. The chance of unnoticed mutations in itself precludes any reliable prediction of the population's outcome. As I say elsewhere in this book, theory holds true so long as it does not have to face reality, whose complexity is overwhelming.

Mutability far exceeds, in the light of the latest discoveries of molecular biology, the numbers regarded as maximal thirty years ago. We could almost say that every gene mutates and does so frequently;

but mutations of very small magnitude (the "neutral mutations" of Goodman and other molecular biologists), and phenotypically unobservable, are by far the most common, as protein analysis has made us realize (hemoglobin, etc.). Behind a façade of stability and constancy, the living world is truly, as Montaigne called it, an eternal seesaw. Stability in variation is the seeming paradox by which all living things are governed. Variability affirms and emphasizes individuality, endowing every creature with its own particular structure and chemistry. Each individual has its own proteins, which make up its personality. Fluctuation, because of the tiny errors in replication found everywhere in the products of the genes, has as a first consequence the "personalization" of every genotype, every phenotype.

The mutations visible by direct observation represent gross copying errors, and mostly create disturbances of the shape and health of the animal concerned. Such mutations are eliminated at the boundary where the monstrous and the pathological begin; this should come as no surprise, for life is incompatible with disorder.

IS COMPETITION UNIVERSAL?

Selection motivated by competition between individuals of the same or different species awards a survival or reproducibility "bonus" to the best endowed. Take the timeworn examples of wolves hunting deer: the fastest runners survive because of being the best fed; they are those who capture the most prey. Lucretius used it first, and Darwin (1859, see 1887, p. 97) took it from him.* There is just one snag: it is not true. Like wild dogs, wolves hunt in packs and run down their prey to exhaustion. The group hunts down the chosen victim, and *all* the members of the pack follow the chase and are in at the kill. There are no champions, who alone appropriate all the food. The solitary hunt is the exception; it is the act of aging males or individuals driven out of the pack, and it does not interest the reproducers, the dominant males. Social ranking order is much more important in deciding the fate of the individual, who may be excluded from breeding (psychological castration).

In some environments and for some species it takes a great deal of imagination to discover selection at work. One example comes to mind: among bush babies or galagos (small lemurlike primates), mainly forest dwellers, there is no trace of competition. The insects they consume in

---

*Ch. Darwin, "Origin of Species," French translation by E. Barbier. Schleicher édit., Paris, p. 97 (examples of natural selection and survival of the fittest).

vast numbers, and all year round, are so plentiful that they have no difficulty in eating their fill. In the case of fruits, things are very different. Their scarcity is due not to overpopulation but to the seasonal cycle of their growth. In some months there may happen to be not a single tree bearing fruit. *Galago elegantulus* is fond of the gums that ooze down the trunks of forest trees, and the supply runs out only for a few weeks in the dry season. Actually, there is never any shortage for these omnivorous lemurs. Predators, chiefly viverrids and birds of prey, eliminate only a small number of individuals. Such elimination is largely unselective, but, as with other predators (e.g., otters), it is doubtless mainly older and sickly individuals that are eaten.

Many cases of selection are age-old but never modify the species. Darwin's example of the wolf and the deer is a case in point; the differences in speed among the individuals making up the population are never eliminated.

Among migrating salmon there is a group too weak to negotiate the rapids or to hurdle the barriers across the rivers up which they have to swim in order to spawn. In both examples the individuals eliminated are born of progenitors who overcame these same obstacles, since only by so doing were they able to breed. The deficient individuals possibly owe their inferiority to the unfavorable conditions in which they developed. In fact, what is eliminated are acquired, not inherited, characteristics. This illustrates the complexity of phenomena concerning the equilibrium of populations and the limited power of selection.

I once studied Orthoptera in various biotopes in France. I never made a single observation enabling me to affirm any perceptible intra- or interspecific competition. Normally, on meadowland, populations undergo numerical fluctuations and are therefore not in precise equilibrium with the environment. But except in rare cases of pullulating swarms [*Locusta migratoria* form *gallica*, *Dociostaurus maroccanus*, *Calliptamus* sp. (*italicus?*)], Acrididae have unlimited food supplies. Outside the breeding period, they ignore one another; only in a few species may individuals assemble by mutual social attraction. Competition is nonexistent. The losses in populations are caused by climatic conditions, predators (which are few, and in some biotopes even negligible), and developmental anomalies as proved by the content of oothecae. By force of circumstances, crippled and diseased individuals, ontogenic failures which may occur because lethal mutated genes ruin the genotype, are eliminated. Similar facts are observed in the suborder of tettigoniids, though with some competition increased by occasional cannibalism (*Tettigonia, Decticus, Homorocoryphus,* etc.). The adult mantis attacks small or medium-sized crickets, and tettigonid larvae, but its casualties are always very light.

In various zoological groups competition may certainly be far from negligible. Sometimes one species hunts another, which is the reason for the rarity of sympatric species, and for there being, on the same territory, a separate ecological niche for each. Sympatric species are not, however, exceptional, a good example being the mixed societies of cercopithecid monkeys in West Africa in which *Cercopithecus nictitans*, *C. cephus*, and *C. diana*, mingled with *Cercocebus albigena*, live practically side by side.

Conversely, the three species of anteater in Gabon, although living in the same place, each have separate ecological niches of their own: The giant anteater (*Manis gigantea*), which lives entirely on the ground, raids chiefly the hypogeal nests of the market-gardening termite; the longtailed anteater (*M. longicaudata*) is a diurnal tree climber that needs to drink water frequently; the common anteater (*M. tricuspis*) is of nocturnal habit and also a tree climber, though less so than *M. longicaudata*. The three species do not interfere with one another. Mutual competition is practically nonexistent. Similar examples can be given for thousands of genera.

Experiments by nonnaturalists demonstrate the role played by interspecific competition in some instances. The rash introduction of the European fox (*Vulpes vulpes*) into Australia has had disastrous results for the native fauna, already damaged by the dingo (*Canis dingo*) brought into the country by immigrant peoples. The mongoose (*Herpestes griseus*) imported into Jamaica and then Martinique to control rats (*Oryzomys antillarum*) and snakes fulfilled its original purpose but went on to attack the indigenous birds and mammals, which suffered heavy losses.

The Seychelles owl (*Gygis alba*) is gradually being replaced by an African owl introduced by man over a century ago to combat rats brought by ships from Nantes and La Rochelle in the sixteenth century. The Malagasy cardinal (or Seychelles canary), imported during the last century, now proliferates at the expense of native grain-eating species of the same size. It would be easy to extend the list of such mishaps.

SELECTION AND DEMOGRAPHY

Demography, which has as its object the quantitative study of populations and their variations, would seem capable *a priori* of showing the influence of selection on a group of living organisms and the ways in which elimination works. It should not be forgotten that the causes of mortality are innumerable and are not always easily known, if indeed they can be at all. In a natural or a laboratory population a characteristic seen in certain individuals can be considered selective with respect to their offspring only if the statistics apply to populations whose environmental conditions do not vary and are clearly known by the observer.

The distribution or strategy of genes, as Waddington calls it, has been minutely analyzed by the most refined mathematical techniques. The literature on population dynamics is enormous; it was thoroughly reviewed in 1969 by Sewell Wright in "Evolution and the Genetics of Populations." This is a highly praiseworthy work, grounded in mathematics but whose worth is precisely that of the theoretical ideas and experimental or other data on which its conclusions are based. This is the reason why mathematics in application to biology has not yet won full acceptance.*

The problem is to determine whether it is true that the dynamics of populations does give an abridged image of biological evolution over a period of time. Although not yet clearly stated, this has been implicitly postulated. To Darwinians, population genetics is the fundamental part of what they call, still in unclear language, *evolutionary genetics.* We do not challenge the accuracy of the statistician's calculations and genetic data, but question whether genetic calculations, equations, and postulates really do concern the evolution revealed by the paleontological evidence and not a mere doctrinaire theory.

Demographic experiments do sometimes occur in nature without human intervention. The biologist must take advantage of these for the lessons to be learned from them. For example, in northern Canada fur hunting is intensively pursued. The Hudson's Bay Company buys skins from the trappers and keeps accurate records. Its account books show that Artic hares (*Lepus arcticus*) and lynx (*Felis (Lynx) canadensis*) pass through cycles of abundance, reaching a peak every nine years, give or take one year. The highest figures are 70,000–150,000 hare skins and 50,000–70,000 lynx. The corresponding low figures are about 15,000.

The lynx feeds on hares and is a fierce predator; both high and low figures apply to both populations, the hunter and the hunted. The struggle is unremitting, as the statistics prove. The evolutionary effect is nonexistent. Morphologically and physiologically, both hare and lynx remain unchanged. The same observation can be made on migratory

*Except for certain kinds of statistical calculations.

animals (migratory insects other than Lepidoptera die without leaving any durable progeny); death shows no differentiation among them; the action of selection is accompanied by no perceptible variation.

In a balanced population, a certain genotype, more than any other, ensures equilibrium. Consequently, all individuals of a different genotype are rather quickly eliminated in accordance with the pressure of selection on the population (we know that this corollary is not experimentally verified; Teissier, 1962).

If the environment alters, another genotype will be best suited to the changed conditions and replace the old one, now inadequate. Of course, this can only occur when selection has allowed varied combinations of genes to persist in the population: *in other words, when selection has not been severe.* This is what we observe in reality. Teissier (1962) stated this in terms of Darwinian genetics: "A very important fact to be noted for the general theory of evolution is that the 'genic fluctuation' of our experimental populations (*Drosophila*) still remains quite large despite the severity toward them of natural selection, and even after much artificial selection remains effective." I have checked this assertion. A 35-year-old population of *Gryllus domesticus* bred in the laboratory and subjected to experiments on group effects proved to be exceedingly heterogeneous, giving widely variant answers both collectively and individually. The fact is this: Animal or plant populations in nature remain heterogeneous, even though subject to natural selection.

Insect or bacterial populations wiped out by antibiotics or poison are an extreme case: Should the protective gene be present, its carrier survives; if not, the individual dies. If these alternatives were repeated for several genes, or alleles of them, there a swift and disorderly change in the population would follow.

In fact such changes seldom occur, since the rule of variation is not an all-or-nothing one. Selection is not severe and the advantage given by certan genotypes is small, even minute. This amounts to saying that the replacement in a population of a previously adequate genotype by a new preadapted one comes about slowly, according to a selection coefficient which may be very small (for calculations, see Sewall Wright (1969) and other geneticists). It should be noted that European populations of the woodland snail (*Cepaea nemoralis*) have been heterogeneous in respect to their shell coloring for at least a million years.

Selection tends to eliminate the causes of a population's heterogeneity and thus to produce a uniform genotype. *It acts more to conserve the inheritance of the species than to transform it.* Thus, these are largely theoretical speculations, for natural populations are highly heterogeneous, being mainly or entirely composed of heterozygotic individuals. It may be added that uniform external appearances often mask a deep-rooted heterogeneity.

The presence, in the same population, of numerous heterozygotic genotypes is attributable either to the weakness of selection, or the neutral or indifferent state of the characteristics determined by the various alleles. Both possible causes often act together and they insure the persistence of the diverse genotypes. There is no need to resort to calculations. The reason why natural populations prove on examination to be so highly heterozygotic is that selection works efficiently only against extremely harmful, pathogenic genes.

Although the theory states that in any population a precise combination of genes offers to the individual a particularly exact adaptation to his surroundings, some of its alleles are so little subject to selection that they survive in succeeding generations (discounting elimination by genetic drift).

This naturally occurring state implies tolerance by selection of the genetic composition of the individuals. If selection operated strictly, all but one genotype—the one insuring maximum adaptation—would be eliminated. In the case of preadapted genotypes (by resistance to antibiotics or pesticides), the opposing gene did not, before coming into contact with the poisonous drug, undergo selection but remained neutral. Is that why it was still present in the population? We may well wonder.

In various regions following heavy treatment with DDT or other pesticides, it has been reported that sensitive populations of flies and mosquitoes have yielded resistant populations. This is a very serious problem for hygienists trying to destroy these Diptera. Mosquito populations (*Culex pipiens*) in the region of Lyons, at first resistant to DDT, have after several applications again become sensitive to it, in some cases after 33 generations (Roman and Pichot, 1972); in Haiti, populations of fasciated *Stegomyia (Aedes) aegypti* that had become resistant to chlorinated insecticides through repeated applications recovered sensitivity to them after 150 generations (Callot, 1958). Similar facts have been cited in the case of the housefly (*Musca domestica*), in which sensitivity recurs after 10–15 generations; as with resistance, it was probably due to preadapted mutants.*

Such considerations, based both on the theory and on the facts, would

---

*To be quite objective, this interpretation is not accepted by all biologists; some attribute the facts (resistance, renewed sensitivity) to extrachromosomal factors (Cochran *et al.*, 1952; Ramade, 1967) and not preadapted mutants.

cause one to accept that in a natural population there are neutral genes or alleles (neither good nor bad for the individual or for the species) which do not undergo selection and form a reserve from which, in the event of environmental changes, the genotype is able to adjust to the new conditions.

To convince oneself, it is enough to consider the infinite number of genetic combinations for the human race. Selection eliminates the weaker, serious hereditary defects, grave deficiencies, etc. In wealthy, advanced societies, the conservation of lethal genes, thus weakening the genotype, testifies to the efficiency of medicine which enables the afflicted to survive and procreate. Such progress turns against man, deteriorates his natural inheritance, and jeopardizes the future of the species in the long run.

Natural selection acts as regulator of the genotype, performing a function of genetic hygiene. As to its role as effective agent of evolution, this is not certain. In fact, if it had the full power attributed to it, it would soon stop evolution. Every noncarrier of the environmentally adjusted genotype would be eliminated. In the event of a change of environment there would be no preadapted genotype to cope with the altered conditions. Thus, natural selection is possible for a population only if it is not too severe. But this is obvious.

The evolution of a zoological or botanical group is not just a sum total of defensive blows to counter aggressions against populations (by the use of a new drug, a sudden and abrupt variation in a physical parameter, etc.). It involves the acquisition of innovations mutually coordinated and precisely adjusted to the older parts of the organism. It is the advance of a whole animal or plant group *toward a given form,* the adoption of a given way of life.

ARTIFICIAL SELECTION BY MAN (CULTIVATION AND DOMESTICATION)

Cultivation of useful or ornamental plants, and domestication of birds and mammals, are a perfect in-depth test of the mutability of the species concerned. Do they thus represent an example of man-controlled evolution?

From the surrounding flora, man chooses, isolates, and promotes for his own personal use a few plants in which he discovers properties that suit his needs—high yield of grains, or fruit that is edible. From the fauna he likewise chooses a small number of animals for their docility and the quantity of meat they supply. In an attempt to improve on nature, he has tried everything to enhance the properties of his crops and domestic livestock. As soon as Neolithic man began to domesticate animals and grow crops,* he practised selection, which he empirically discovered to be successful.†

The difference between natural and artificial selection is in the end purpose. Natural selection operates for the greatest good of the species, artificial selection for that of man. They work upon the same raw material: mutations. In artificial selection the choice of sires is the work of man's hand; in natural selection the fittest to ensure reproduction of the species are those death has spared.

Artificial selection creates nothing by itself. The same is true for natural selection. It sorts out or gathers together broodstock (in the case of multiple genes or alleles scattered through a population). In this manner it has increased the butterfat content of cow's milk, the saccharose content of the sugar beet, the length of the cotton fiber, the length and softness of sheep's fleece, and so on.

Several major mutations in the dog, the pigeon, the ox, the sheep, the rabbit, the silkworm, the bee, wheat, barley, maize, fruit trees, roses, snapdragons, tobacco plants, etc., have been exploited on a grand scale. Mutations incompatible with life have been eliminated, as they have been under natural conditions; but man preserves teratological variations having a usefulness for him (hornless sheep, ancon sheep) or fanciful character (bulldogs, hairless dogs, goldfish, telescope goldfish, Siamese fighting fish).

According to Darwinian teachings, domestication is a speeded-up evolution controlled by man. Speeded-up, because mutants considered to be worthwhile are immediately retained and are thus have an advantage with respect to wild types. They correspond to the carriers of the privileged genotype in natural selection.

The products of domestication deviate more or less from those found in free nature and sometimes border on monstrosity. It is hard to visualize how lapdogs, Yorkshire terriers, or Pekinese could survive in the wild. Certainly they would not last long in the woodlands or pastures of our temperate zone. The albino rat, albino angora rabbit, and how many other domesticated animals would, if set free in the countryside, perish in only a few days.

Often the domestic form resembles its wild ancestors. For example,

---

*It is claimed, but not proved, by prehistorians that the domestication or, rather, conditioning of certain animals was started in the upper Paleolithic.

†This is not a simple supposition; several species of domesticated dog lived in the Neolithic.

the farm turkey differs very little from that of the forests of Yucatan or the southwestern United States; the domestic guinea fowl is almost identical to the one that lives in flocks in the African grasslands. Both these gallinaceous birds have mutated much less than the farmyard fowl. Yet domesticity of the turkey goes right back to antiquity since they were bred by the Mayans and Aztecs in their villages (they had only two domestic animals, the dog and the turkey). The high mutability of the fowl is evidenced by its numerous breeds and freaks, which yet remain very close to the ancestral stock, itself subdivided into three geographical subspecies, *Gallus gallus gallus* of the Burma jungle, *G. g. murghi* of southern Kashmir, and *G. g. bankiva* of Java and Bali.

Among the very numerous races of the domestic pigeon, some, e.g., pouter pigeons, peacocks, the "carneau" of northern France, deviate widely from their ancestor the rock pigeon (*Columba livia*) that still lives in a wild state in rocky places, whereas some, the carrier pigeon being an example, are very close to it. There is no mixological barrier between these breeds although some date from thousands of years ago.*

The dog is probably the earliest domesticated animal, no doubt because of its ability for keeping man company. The dog of Senckenberg (near Frankfurt on the Main), whose bones were found mixed up with those of an aurochs, lived 9000 years ago (dated by the pollens found with it) and may have been tame. Earlier still, at Jericho (10,800 ± 180 years old, ^{14}C dating) and in Persia (Belt cave) (11,480 years old), there lived dogs that were in all likelihood domesticated. It is presumed that the dog became a companion of man toward the end of the Paleolothic, at the time of the last Ice Age.

The Neolithic farmers had several separate breeds, the commonest slightly smaller than the average for present-day dogs, another smaller still, and a third definitely larger. From the Neolithic to the Iron Age four main breeds of dog existed in Europe: *Canis familiaris inostranzewi*, which is thought to have produced the Eskimo dogs bred for pulling sleds; *C. f. matrisoptimae*, from which our sheepdogs are probably descended; *C. f. intermedius*, to which various breeds of poodle, etc., are linked; and *C. f. palustris*, the small dog of the Neolithic lake-dwellers, from which Pomeranians and terriers are derived.

It is possible that the breeds of succeeding ages may not all have the same origin. Some zoologists assign as the ancestor of North African dogs the Egyptian jackal, *Canis lupaster*, but this is by no means certain.

---

*The domestication of pigeons goes far back into the distant past. It was practiced in Mesopotamia some 6000 years ago, because clay figurines of pigeons have been found (Halaf period).

The Indian pariah dogs (*C. familiaris indicus*) that live ferally around urban agglomerations, and their ally the Australian dingo (*C. f. dingo*), seem according to recent studies to differ only slightly from their ancestors, the stock which produced the domestic dog. The opinion that the dog is a modified wolf still has some supporters, but zoologically speaking there are weighty arguments against it. The origin of the greyhound remains obscure; this Mediterranean breed may descend from one particular dog, but all its characteristics seem to indicate that it belongs to the species *Canis familiaris*.

The range of dog breeds is even wider because selection has taken place in different ways, according to esthetic value or oddity, aptitudes for hunting, and use as watchdogs. Thus, a great many mutations more harmful than useful to the species have been maintained. The biggest variations are in size (King Charles spaniel, 0.75 kg; mastiff, 50+ kg), skull shape (bulldogs and mops, with shortened upper jaw and a more or less deformed mandible projecting beyond it), length of legs, ears, coloring, length of body, and even hairlessness. All these breeds, however different from one another, cross-breed readily without any loss of fertility. They are less separated from one another than geographical races of certain wild species, e.g., those of the North American frog *Rana pipiens* [the spawn of the race found in New England fertilized by sperm of the Florida race produces abortive embryos or abnormal tadpoles (Moore, 1946, 1949)], or *Drosophila paulistorum* (Dobzhansky, 1962), which is found from Sao Paolo to Mexico. Of course, inequality of size may in some cases make cross-breeding impracticable; for example, a mastiff cannot service a King Charles spaniel, and vice versa!

From all this it is quite clear that dogs selected and kept by man in a domesticated state *remain within the boundaries of the species*. Tame animals that have reverted to the wild state lose the characteristics produced by mutations and fairly quickly resume the original wild type. They get rid of the man-selected characteristics. This demonstrates, as we knew before, that artificial and natural selection do not work in the same way.

Hutch-bred rabbits introduced into Australia by Governor Philipp in 1788 found an environment favorable to their proliferation, and they now number hundreds of millions. The important fact for us is that they look exactly like wild rabbits.

The wild goats of Juan Fernandez Island are a common type. In North America the horses that have gone wild or semiwild have kept some of the characteristics of their respective strains: height, dappled hide, high-ridged faces.

The ferret (*Putorius putorius furo*) does not persist in the natural state; the numerous escapees have not founded a stock. Incidentally, there has never been a single case in any natural animal population of an albino mutant prevailing over the normal type.

Eskimo dogs that have gone wild retain their ancestral type, perfectly adapted to the Arctic climate. In this case artificial selection worked in the same direction as the natural one it had displaced, so it is little wonder that the strain should continue with no human intervention.

What genetics teaches is that heterozygotic populations in the natural environment lose the alleles of the wild type genes owing to Mendel's laws of segregation of characteristics and subsequent genetic drift, together with the selection they undergo (elimination of genotypes with mutated characteristics, unfavorable in the natural state). If they are homozygotic with respect to their racial characteristics, the alternative is three-pronged: Either they persist as such, or they die incapable of coping with wild life, or they undergo *reverse mutations*, an eventuality which seldom occurs but which brings them back to their previous state.

Such mutations have been observed in *Escherichia coli* by Lederberg and Tatum (1946): out of cultures of mutants incapable of synthesizing, for example, the amino acids threonine and leucine and a vitamin, thiamin, have appeared, with a frequency of 10^{-6} to 10^{-8}, bacilli that are able to do so. Two kinds of mutations are distinguished in *E. coli*, those directly involving the gene (same locus in the DNA molecule), and those due to the intervention of suppressor genes located at another site.

The diversity and numbers of breeds of domestic animals or cultivated crops are increased by cross-breeding as frequently practised by stock breeders and farmers; the combination of characteristics by hybridization and selection increases the number of genotypes and the variability of the species. The changes brought about in the genetic stock affect appearances much more than fundamental structures and functions. In spite of the intense pressure applied by artificial selection (eliminating any parent not answering the criterion of choice) over whole millennia, no new species are born. A comparative study of sera, hemoglobins, blood proteins, interfertility, etc., proves that the strains remain within the same specific definition. This is not a matter of opinion or subjective classification, but a measurable reality. The fact is that selection gives tangible form to and gathers together all the varieties a genome is capable of producing, but it does not constitute an innovative evolutionary process.

Ten thousand years of mutations, cross-breeding, and selection have mixed the inheritance of the canine species in innumerable ways without its losing its chemical and cytological unity. The same is observed of *all* domestic animals: the ox (at least 4000 years old), the fowl (4000 years), the sheep (6000 years), etc.

Despite the magnitude of variations, several of which border on the monstrous, experimental domestication–selection does not result in the creation of the new species. Races do not keep to themselves and are able to hybridize with one another without loss or decrease of fertility. Domestication and cultivation demonstrate the rather narrow limits within which the species can vary without danger, but do not give it any evolutionary motion.

Natural selection preserves individuals that are physiologically in equilibrium with the environment; thus, it maintains a certain specific type. Selection practised by man has quite a different effect. One example will suffice: The genotypes of a dog (*Canis familiaris*) and a jackal (*C. aureus*) are extremely similar and, with a very few differences, subject to the same mutations. But the jackal species is highly stable whereas the dog species is divided into numerous races and subraces. The former is subject to natural selection, which eliminates variants and stabilizes the species, the latter to artificial selection, which preserves the "abnormal" and facilitates their survival.

ERRORS OR INABILITIES OF SELECTION

We have already pointed out that by exaggerating certain characteristics to an extreme, hypertely destroys the balance of the anatomy and functions of a living creature to such an extent as to create a real threat of extinction for the species. The terminal forms of hypertelic lines of descent are, however they may be interpreted, the result of a long evolution during which selection (assuming for the time being that its part in the process is an active one) has never ceased to operate, always in the same direction.

But apart from major hypertely, anyone, however little of a naturalist he may be, can discover a great number of characteristics which, far from being beneficial, are harmful to the individual. For example, the deer has such a marvelous display of antlers that the French poet La Fontaine wrote of him: "His antlers, a dangerous adornment continually delaying him, hinder the service rendered by his hooves, on which his life depends."*

Speed is of little use to this herbivorous animal for if discovered when browsing or in his lair, he is attacked before he can get away safely. He is

*Son bois, dommageable ornement,
  L'arrêtant à chaque moment,
  Nuit à l'office que lui rendent
  Ses pieds de qui ses jours dépendent.
          (*La Fontaine, Fables,* VI, 9)

his own worst enemy. As proof of this, the Cervidae and Bovidae have during millions of years of evolution acquired cutaneous glands whose pungent secretions mark the individual's territory, the paths followed by the herd, and reveal the presence of the male. The most specialized marker organs are the tear ducts, suborbital slits into which the preorbital glands pour their secretion.

For our purpose the most interesting are the tegumentary glands of the hooves, called, according to their site, carpal, tarsal, metatarsal, and interdigital glands. Their unctuous and scented secretions lay a rather durable trail of scent which is used by the members of the herd to keep in touch with one another and recognize their territory; but they also enable big game animals (wolves, wild dogs, the great cats) which are all macrosmatic, to track down their prey who so naively give themselves away.

The ruminants have undergone a twofold evolution, one making them run faster (changes in the leg, relative proportions of bony segments, hooves) and the other relating to social or sexual relations and chemical signals, making them more vulnerable to their natural foes. Both evolutions seem to have been synchronic and imply a selection working in two distinct directions. We might add that males of the goat (with subcaudal glands below the tip of the tail), the deer (*Cervus elaphus*), the ram (*Ovis aries*), the llama, etc., give off a powerful scent that indicates their presence a long way off and thus warns their enemies.

From a logical viewpoint, i.e., an anthropomorphic one, a selection favoring the herbivores should have made them scentless so that macrosmatic big game animals could not track them. Evolution did quite the opposite, and the herbivores are still there.

There is no end to selective aberrations. Bush babies or galagos put their hands, moistened with urine, to their perches, thus indicating their presence to their fellow creatures, which is understandable given the social proclivities of these lemurs, but also giving them away to predators such as the tree-dwelling Viverridae (*Poaina,* polecats, and the like). Black rats (*Rattus rattus*) splash their usual trails with tiny drops of urine, which enable terriers to find them easily, and so on. How true it is that you cannot think of everything!

If selection has preserved such aberrations, we have to consider it misguided. But our reasoning is at fault. A Darwinian would retort that between two alternatives, difficulty for the sexes in finding one another and attack by predators, selection chose the former. We shall continue to think it a poor choice, and one which makes the species highly vulnerable.

It is hard to see why some structures should maintain the peculiarities of their selection. The coelacanth (*Latimeria*), an authentic relict of the Permotriassic fauna, has a wealth of paradoxical characteristics so unexpected and unpredictable that Cuvier, for all his genius, could never have deduced their existence from merely looking at its skeleton. The lung is an adipose mass attached to the ventral face of the esophagus. The kidneys, partly postcloacal, and the major sympathetic nervous system have migrated from the dorsal region and come to rest on the ventral floor of the abdomen. The minuscule brain is surrounded by thick adipose tissue that fills the enormous skull cavity. The eggs, as big as oranges, are shell-less and more voluminous than those of the biggest oviparous sharks! It is absolutely impossible to see what role selection can have played in the acquisition and conservation of such unlikely characteristics.

NATURAL SELECTION, OR FINALITY IN ACTION

For Darwinians, following their founding father, evolution is based upon the survival of the fittest, that is to say those which are best endowed and which bear characteristics compatible with life in a given environment. Assigning to natural selection the effective execution of evolution means explicitly and implicitly attributing to it a meaning and end. When Darwinians argue that the finality observed in biological phenomena is an illusion, a deceptive appearance, they forget or fail to recognize the very foundation of their interpretation of nature, one heavy with philosophical consequences. By making selection the agent of the evolution of the fittest, they confer on every living thing an inherent finality that becomes the supreme law of the individual, the population, and the species. When is Darwinian doctrine going to be subjected to a thorough, critical reevaluation?

To say that natural selection is not directed because you cannot see anyone or anything directing it is not an acceptable objection. Indeed, if selection is unintentional, it is found to be random. Darwinians and their molecular biologist disciples conceive of it as making use of aleatory material for the benefit of the species; thus, it sorts out and directs. It would be a strange intellectual position to replace one chance by another one. But we can be quite formal; selection is not a random phenomenon. It is by its very nature a teleological one.

Since no one can be seen directing it, the Darwinians think this sufficient proof to declare it to be undirected. *This is a very grave philosophical and truly anthropomorphic error*, making selection as an active and transcendental entity.

How is it possible, in that case, to speak of pseudoteleology, or teleonomy (which, etymologically, means the laws of purpose)? To factual
or immanent finality the Darwinians add a finality of a higher order,
inherent in life and constantly active; in the biosphere it presents the
characteristics of a transcendental finality. No biological or philosophical
system has gone farther toward the finalization of living creatures.

The selective act is inseparable from an end, whether directed by man
in the case of artificial selection, or in the case of natural selection by
death—death which never strikes at random, but, on the contrary is an
efficient agent in natural selection. In reality, the very purpose of selection is to finalize life.

Lamarck attributes to the organism itself the faculty of being its own
adaptor to the external environment. Here, too, we confront an interpretation invoking a finalizing mechanism. When we venture into the living
world, as we must, the immanent finality shows itself in practically all
structures or functions and control mechanisms, and natural selection
and the capacity for adaptation are seen as the agents of a finality of a
transcendental type. To eradicate finality from biology is a vain attempt
because it goes against reality, and those who try to do it are inspired by
philosophical theories or theses which act as blinders to the facts.

CONCLUSION

Selection in nature acts upon species to eliminate the "not-so-good,"
the flawed, the disabled. That is its chief role.

Interspecies competition plays a part in the spatial distribution of
populations and species. It causes some evictions, and limits the number
of sympatric species. Usually it is a matter of local equilibria. The outcome of the struggle is not the same everywhere; in one place the species
is rejected from a biotope, in another it is victorious and prosperous.
Consequently, the smaller the territory they inhabit, the more species
are threatened in their existence. The fauna of small islands is extremely
sensitive to the introduction of alien species. The apterous birds (Ral-
lidae, columbiforms) which flourished in the archipelagos of the Indian
Ocean before they were populated by man have not withstood predators
brought from Europe. But man has massacred so many insular species
that it is hard to know what share of this destruction* was caused by
other predators.

---

*For an account of the annihilation of the large wingless birds of the Indian Ocean
islands, see Strickland and Melville (1848). The dodo of Mauritius (*Raphus cucullatus* L.),
that of Réunion (*Raphus solitarius,* Sel.L.), and the solitaire of Rodriguez island (*Pezophaps*

The facts relating in one way or another to selection are of much greater concern to the populating of our planet than to biological evolution. It is an error to confuse the two things. The "evolution in action" of J. Huxley and other biologists is simply the observation of demographic facts, local fluctuations of genotypes, geographical distributions. Often the species concerned have remained practically unchanged for hundreds of centuries! Fluctuation as the result of circumstances, with prior modification of the genome, does not imply evolution, and we have tangible proof of this in many panchronic species: cockroaches, poduriform collembolans, Hyracoidea, bacteria, etc.

The genic differences noted between separate populations of the same species that are so often presented as proof of ongoing evolution are, above all, a case of the adjustment of a population to its habitat and of the effects of genetic drift. The fruitfly (*Drosophila melanogaster*), the favorite pet insect of the geneticists, whose geographical, biotopical, urban, and rural genotypes are now known inside out, seems not to have changed since the remotest times.

---

solitarius) were destroyed by the Europeans who colonized these islands. Lord Rothschild (1907), in his *Extinct Birds*, gives great number of facts about vanished species and the causes of their extinction.

VI

Evolution and Adaptation

GENERAL REMARKS

The concept of adaptation is imprecise because it is not easy to measure and quantify the relationship between a structure and its function, or between a function and the environment.

Whenever biologists and philosophers judge the extent of an adaptation on the basis of our logic and readily formulate value judgments, they are guilty of methodological carelessness; they fail to realize how anthropomorphic their method really is.

The childlike finalism of Bernardin de Saint-Pierre, who discerned everywhere the intervention of divine Providence, has harmed the case for adaptation. In fact a great many naturalists have been just as finalistic, though with more discretion. The partisans of Darwin, because of their desire to discern a function and use for an animal's every organ and every act, have shown themselves to be just as naïve as this eighteenth century French writer; for it must not be forgotten that the thesis of omnipresence of adaptation has as its corollary that every organ, every physiological mechanism, performs some favorable act for the individual or the perpetuation of the species. Such is the belief of the modern Darwinians, from T. H. Morgan to E. B. Ford. For them adaptation is everywhere and without it life cannot flourish. No unadapted creatures could exist since they would not be viable, though it should be remembered that each living being is a complex whose components are not all equally suited for the purpose they serve. To the idea of perfection we prefer that of compatibility with life since the greater relativity of the term corresponds more closely to reality.

Every living thing is a physicochemical complex having the ability to

131

maintain itself in equilibrium with its environment, so long as the latter does not vary too greatly. The range of physicochemical conditions compatible with life is a narrow one.

Adaptability appears to be a property the living creature cannot do without. Whether we call it homeostasis or self-regulation, what counts is recognition of its role. It depends upon systems mutually adjusting themselves, so that despite environmental changes the economy of the organism is unaltered.

There are a great many self-regulating systems which operate at various levels of organization: molecules, cells, organs, organic mechanisms. Whereas in every case the materials used are ions, atoms, molecules, these nevertheless work their effects on parts of higher rank. Molecular biology with its repressors and structural genes describes regulating systems at a genic level: hormones stimulate genes and finally work on the organs, using enzymes as their intermediaries.

The biochemical systems to which we refer, and which participate as units in the normal life of the cell and organism considered, are not the only regulators the animal has. Defense against foreign proteins introduced into the organism by a parasite, virus, bacteria, unicellular or multicellular animal, or fungus consists of a set of reactions or an immunity which in many cases succeeds in destroying the parasite or neutralizing its pathogenic action.

The regulating systems also involve organs, and even substantial portions of the animal. The eggs of sea urchins and certain hydroids, etc., sliced into two or even four, in planes passing through the polar axis of the animal or plant, produce as many normal and complete embryos as there are fragments. We call such eggs *totally regulative* eggs.

The regeneration of missing parts is also a form of regulation. Unevenly distributed throughout the two kingdoms, its frequency and potency are incomparably smaller among animals than among plants, where it very commonly develops into asexual multiplication.

The adaptation of major functions to the environmental conditions by means of control mechanisms is universal, it is a *sine qua non* for survival, whereas adaptations to particular conditions are irregularly distributed and not equally efficient.

For example, euryhaline fish species, able to live in waters of uneven salinity, are in the minority, having a defense mechanism enabling them to pass unharmed from a hypotonic environment (fresh water) to a hypertonic one (sea water) with respect to their internal environment (the case of eels whose cutaneous mucus insulates them from the outside environment), or a mechanism inhibiting the excretion of their own salts (salmon and other potamodromous fish). The immense majority of fish species are strictly bound to a given salinity (stenohaline species) ranging from fresh to sea water, including the brackish water to which a few fishes are adapted.

The consideration of survial rates of individuals as a function of certain characteristics, as is done in estimating selectivity coefficients, is not without its value, but adaptation is often too subtle to show up in the demographic tables for a given species since other more forceful factors obscure or nullify its usefulness. The near impossibility of judging the practical value of adaptation accounts for the many disagreements that have arisen among biologists on the subject.

Despite the legitimate reservations concerning the usefulness somewhat frivolously attributed to certain organs, functions, and behaviors, it would be absurd to cast doubts on the reality and extreme importance of the organic as well as functional balances between the living creature, its environment, and its mode of existence.

RELATION OF ADAPTATION TO SELECTION

Any theory of evolution has to explain, in addition to the oriented transformations of living things, the above-mentioned balances that have been called at one time or another adjustment, harmony, adaptation.

Lamarck and Darwin, the founding fathers, each in his own way set forth an explanation of adaptation. According to the former, the living creature reacts to the actions of its environment upon itself. In a physiological, psychic, or organic way it responds to the aggression of the stimulus from the outer world. Its response thus enables it to survive.

Darwin and his disciples claim that selection and adaptation go hand in hand, the latter resulting from the former, which *harmonizes* the creature's relations with its environment. Conceived in this way, selection fulfills the functions and takes the place assigned by Bernardin de Saint-Pierre, in his anthropocentric system, to what he calls "Nature or Providence." Selection, the Darwinian's anti-chance, works for the greater good of the species. The good Bernardin de Saint-Pierre (1784, see 1823, p. 334) does not go quite so far: he merely notes that "Nothing has ever been done in vain"; it is a less compromising claim.

The organized being takes the forms of a system which, *obligatorily* and *constantly*, remains in equilibrium with its outside environment. Not only does it create negentropy but, by self-regulating devices, it arrives at a compromise, a *modus vivendi* with the environment. The balance takes place all the time, a single word of very wide scope expresses this perfectly: regulation. It involves all the activities of living things. Some examples follow.

An animal's behavior is subject to the most watchful control. Messages from the outside world, picked up by its sense organs, are transmitted to the nervous system, which processes and integrates them and then triggers off automatic and *appropriate* reactions which place the organism in conditions compatible with life. Sensation is both the informer and, in Henri Piéron's phrase, "the guide of life"; this is as true of the ameba as it is of man.

The controlling reaction is a fundamental though not easily quantifiable characteristic of the epigenetic phenomenon that we call life. The physicochemical factors (heat, moisture, atmospheric ionization, chemical substances diluted in air or in water) operate necessarily upon the whole of the living creature; a change in them affects metabolism as well as behavior.

Let us take as an example the "defensive" immunizing reactions set up by the introduction into the organism of alien proteins, whether produced by internal parasites or ingested with food. The animal thus attacked takes defensive action by generating, by means of the cells of the reticuloendothelial system and the white corpuscles of the bloodstream, antibodies that combine with the intruder substances and make them inoffensive. This is a purely chemical self-adaptation reaction.

Inside the organism the mobile leukocyte is activated by substances dissolved in the blood plasma or the intercellular fluid that comes into contact with it; these substances exert attraction on it, induce it to creep, and direct its motion. This cellular behavior is totally automatic. It is hard not to see in it a finalized process, without lapsing into anthropormorphism. It is tempting to say that leukocytes that gorge themselves on pathogenic microbes defend the organism like citizens springing to arms in defense of the city. At all events, this is how it is stated in the loftiest classic handbooks and texts.

Darwinians argue that such defense mechanisms have been gradually organized by successive selections. If this be true, selection does act as a transcendental and prophetic agency of biological finality.

But selection is not an entity in itself, and we cannot endow it with discrimination. Its working instrument, we know, is death: death that eliminates the flawed, the deficient of all kinds, the unadapted, and the weak. The creature dies because it is at odds with its environment. It all happens as if death acted for a purpose greater than itself but the reality of which is entered in the demographer's tables. As a differentiating agent it acts as a policeman, maintaining biological order, punishing every misdemeanor and negligence that endangers it. By the same token, it safeguards populations, both plant and animal, protects the species against deviations that place its future at risk; however paradoxical it may seem, death saves the species by keeping it pure and healthy.

Demography teaches that in nature losses are seldom, if ever, light. In a stable population one pair, and only one succeeds another. Every other individual disappears before being able to procreate. Hence, death plays the leading role in the exercise of selection, which does, however, have a few effective agents of its own, the chief of which is *differential fertility*. In an unstable and heterogeneous population containing two varieties or species living side by side, the one that is the more fertile (after a time that can be calculated from the numbers recorded) prevails over the other, assuming, of course, that both have the same coefficient of selectivity.

In the end, we see that adaptation depends on the use of widely varied means, some inherent in the genotype of the species, others in the environment.

THE LIMITS OF ADAPTATION

Considerable importance is attached to the ability of all living things to adapt to circumstances, to their environment. Yet such adaptability is limited and is soon exhausted in the event of the environment passing a certain level of change, such as seen in the enormous losses suffered by the species from external causes. Lacking the appropriate material, selection proves incapable of endowing a population with properties that would establish it in equilibrium with the environment. Infectious diseases emphasize the deficiencies of adaptation; the immunizing mechanisms are powerless to curb the advances of parasite, bacteria, or virus.

The limits of adaptation may be determined by subjecting the organism to far-reaching environmental changes. Thus, few multicellular animals can survive the transition from an aerobic to an anaerobic existence; by chance a handful of insects. Helminths are normally anaerobic; free oxygen is poison to them. Pulmonary respiration in reptiles, birds, and mammals excludes all other kinds, whereas urodele amphibians breathe more through their skin, when it is wet, than through their lungs, which can even be removed without causing death. Conversely, in a dry atmosphere their skin hardens and fails to allow gaseous exchanges, so the animal dies. The Plethodontidae, all found in North

America except the European genus *Hydromantis*, are normally ap-neumonous; they live in water or very damp places and the skin alone effects respiration. *Aneides lugubris* breathes mainly through the digits, with their blood sinuses spread underneath the skin.

Adaptation fairly often goes beyond a mere unity between structure and function. In such cases, it produces its effects by complex and subtle means. We are referring to the way in which certain flatworms and opisthobranchiate gastropods make use of weapons they take from their victims.

Planarians of the genus *Microstomum* (archeophorus Turbellaria), whose numerous species live some in fresh, and others in sea water, prey on creatures such as Hydroids, whose bodies have an abundance of urticating cells containing a very active poison. The hydras are de-voured and digested until their cnidoblasts (stinging organs of the ur-ticating cells), left intact, are set free in the worm's intestine. These tiny organs are then captured by pseudopods and sent by the cells in the wall of the intestine into the cavity of the gut. The intestinal cells expel them into the mesenchymatous tissue filling the intervisceral spaces. There, they are phagocytosed by mobile cells which carry them as far as the tegument, composed of ciliated and glandular cells. They are wedged in the tegumental cells in such a way that their tips are turned outward and form projections. A mechanical stimulus excitation applied to this tip causes the nematocysts to discharge [Meixner (1923) maintained that they did not burst, but Kepner and Nuttycombe (1929, 1938) have shown them to be perfectly functional].

When the *Microstomum* has used up all its ammunition, it eats more hydras and so gets a fresh supply of cnidocysts that distribute them-selves evenly over the surface of its body; when there are enough of them the worm calls off its attacks on hydras, which would appear not to be its favorite prey.

Most aeolian opisthobranch gastropod mollusks have long papillae on their dorsal face. These hollow papillae, into which an extension of the digestive gland penetrates, do or do not open outward. The cavity of the papilla fills up with nematocysts which, as with *Microstomum*, come from hydroids over which the *Aeolis* crawls, browsing upon their sur-face. The urticating systems slip between the cells of the papillary walls, undergoing a slight degeneration in the process. They turn into cnidoblasts which, when mature, break off from the wall, remaining inside the papillary cavity. When an enemy threatens the *Aeolis* the wall contracts and shoots them out. In species with closed papillae on the outside, the papilla apparently detaches itself and frees cnidocysts (Burgin-Wyss, 1961).

These cellular behaviors, staged and executed with extreme precision, show the perfection that adaptation may attain and its close relationship to the immanent finality of living beings and their constituent parts.

NOT ALL IS FOR THE BEST IN LIVING THINGS

The very bases of Darwinism—yesterday's or today's—inevitably lead us to attribute to the animal or plant, if not anatomical and functional perfection, at any rate a constant aspiration to it. This idea flourishes in the work of modern Darwinians: T. H. Morgan, R. A. Fisher, E. B. Ford, etc. Julian Huxley has shown great ingenuity in developing and defending the principle of the universality of adaptation.

Like Huxley, I, too, affirm the presence of adaptation and maintain, as he does, that the living creatures cannot do without it. One would have to be blind to fail to see the harmony that reigns in most cases between structure and function, or between both of these and the creature's way of living. But adaptation does not always attain perfection, and organs are often apparently ill-adapted, useless, or even harmful.

Julian Huxley, on p. 478 of his masterwork (*Evolution: The Modern Synthesis*, 1943 edition) heads section 8 in Chapter 8: "Adaptation and Selection Not Necessarily Beneficial to the Species" and declares outright: "So far, we have been discussing adaptation more or less *in vacuo*. . . ." The author seems to have been unhappy in writing this chapter and to have done so out of a sense of his duty as a scientist, but without going quite so far as to criticize his own doctrine.

As examples of defective adaptation Huxley cites only harmless cases, whose significance he reduces, and refers to intermediate semiintra-specific, semiinterspecific selection, intergroup selection, and other ad hoc entities. He compares selection to a god creating the best and the worst indifferently, the sublime and the horrible. To substantiate his interpretation, he writes: "We need only think of the ugliness of the *Sacculina*, or a bladder-worm, the stupidity of a rhinoceros, or a stegosaur, the horror of a female mantis devouring its mate or a brood of ichneumon flies slowly eating out a caterpillar" (Huxley, 1945, p. 485). Huxley's choice of examples does not demonstrate that evolution is defective; they shock our human preconceptions, logical, moral, or esthetic, but our value judgments have nothing to do with evolution.

The living creatures that survive today are the "chosen" of a beneficent antichance; natural selection, whether it has acted brutally or gently, is one. Yet we do find anatomical systems, physiological mechanisms, that are detrimental to the species; in these cases the biologist should look for the "ill-intentioned" cause frustrating natural selection, assuming the latter to be, in fact, omnipresent.

FAULTY EVOLUTION

If selection consciously oversees evolution, how is it possible that, through the ages, so many lines have taken paths which endangered them? Still, this is what has often happened, and what accounts for the numerous families and orders now extinct. Evolution has surely had its failures, and occasional catastrophes.

No extinct line reveals the causes of its disappearance. Paleontology does not restore the conditions in which the organism lived. Nonetheless, we observe that vanished species have often had in common particular highly specialized organs, inordinately developed. One is inclined to attribute to such excesses an important, if not a preponderant, role in the extinction of these species.

The classic example of this phenomenon is that of the Titanotheriidae or Brontotheriidae, which started out as animals as big as a fox and ended up the size of an elephant. As they grew in size they acquired powerful horns and bony protuberances of the skull. With an enormous facial mass and tiny brain cavity, they presented an obvious anatomical disequilibrium. Their existence on the scale of the earth's history was short, approximately less than 30 million years.

A more significant case is that of the Rudista. These bivalve mollusks manufactured shells of preternatural thickness, accumulating carbonate of lime as no other animal has ever done before or since. Their history begins in the Rauracian and ends in the Danian, at the end of the Mesozoic era; it lasted only 76 million years. The Rudista inhabited coral reefs, in warm, transparent waters like those of the South Seas in which are found the biggest bivalves, the giant clam or *Tridacna*, and rainbow-hued corals and fish.

Some paleontologists, searching further back in time, believed the earliest ancestors of the Rudista to be among the preheterodont bivalves of the genus *Megalodon*, from the Devonian to the Triassic, and *Protodiceras* of the Liassic. Their evolution mainly took the form of the simplification of a hinge, already observable in *Diceras*, the oldest indisputable rudistan. "Thus the hypothesis has been advanced that as early as the Devonian a group of preheterodont Lamellibranchia set out on the path leading to the Rudista" (Dechaseaux, 1960, p. 2147).

The direct progenitors of the Rudista must have looked like *Diceras*—thick-shelled, subequivalve, of an extended and twisted cone shape like an antelope's horn. It may be taken as accepted that the Rudista are related to the heterodont eulamellibranchiate bivalves. They very soon branched off into two lines, which we are not too sure were homogeneous; in one of them, the richer in families (5, as many small branches) and genera (57 described up to 1949), the mollusk is attached to its support by the right-hand valve, in the other by the left. Without going into the details of their evolution, let us mention some of the more "extravagant" examples of it. In some genera, the right-hand valve is a huge and massive cone up to 50 cm high; the left one serves as operculum or lid, and is much less thick. The hinge consists of a tooth with two fossae on either side on the right-hand valve and one between two teeth on the left. The very long teeth are set in deep depressions. This arrangement does not allow the valves to rotate around the axis of the hinge; the opercular valve can only open and shut vertically.

I am at a loss to understand how this heavy shell should manage to be lifted without a powerful elastic ligament antagonistic to the adductor muscles which, as the name indicates, close the valves by contracting. In all events, the opening of the valves was only a slit just big enough for water to pass into the palliate cavity, bringing the plankton on which the lamellibranch fed (it is agreed that the Rudista had no syphons). The alimentary requirements of the most highly evolved Rudista can only barely have been met, and the slightest shortage of plankton must have threatened their existence. The latest Rudista have been found in Damian deposits of Catalonia, whose continuous and uniform sedimentation since the Senonian rules out any hypothesis of destruction by cataclysm.

Without venturing an explanation of the exact cause of their disappearance, I merely note that for over 70 million years a group of mollusks followed a controlled evolution bounded within exceedingly narrow limits that cannot be considered to be favorable to them because it ended in total extinction. Their story shows us that a misguided evolution with respect to persistence of the line can show itself just as consistent and constantly directed as one more favorable to the species. This is highly significant.

It is very unlikely that, despite changing oceanic conditions, variations of temperature and salinity can have controlled and directed such an evolution. One must conclude that internal factors were operative. We shall return to this point later.

The birth, prodigious apogee, and catastrophic end of the dinosaurs is no less instructive. These reptiles populated every earthly environment from the Triassic to the Damian, when, for unknown reasons, *all* of them vanished. Most of them, especially the terminal species, clearly and by a vast majority had a tendency to generalized hypertely (excessive growth, exceeding its "goal"): gigantism, hypertrophy of the cephalic "appendices" of their armor plate, tiny brain, enormous medullary canals, huge and fragile eggs, etc. But even a Sherlock Holmes needs some evidence to work with, and it is exceedingly difficult, perhaps even a vain hope, to attempt to reconstruct the situation existing 100 million years ago. Everything in our interpretations is disputable.*

For instance, hypertrophy of the medullary canals, which causes reflex actions to predominate over all other behavioral reactions, has been regarded as disadvantageous to the animal since it condemns it to automatism, thus making it ill-adaptable to circumstances. It is equally possible to argue the opposite case, that it is advantageous to the species in that it neutralizes some of the ill effects of gigantism. In any big organism the nerve impulse has a very long path to reach the brain, so that sensory messages arrive with a time lag. Because of their placement, the medullary canals shorten the travel and reaction time; on a reduced scale the same phenomenon occurs in the nervous system of hirudinates, mollusks, crustaceans, and insects, whose ganglia become important reflex-mediating centers, e.g., the metathoracic ganglion of jumping orthoptera, which commands the autonomy of the hind legs. The giant axons allow ultrarapid transmission of nerve impulses. The dinosaurs may have had some too, through which the nervous input traveled more quickly than in small axons.

The physical state of the dinosaur was not so bad after all, since even the most gigantic among them lasted tens of millions of years, not a negligible time on the scale of human life and the history of our globe.

The dinosaurs' evolution was by no means aleatory. It is even possible that the two main orders, Saurischia and Ornithischia did not have the same ancestors; discussion of this point would be out of place here. Each of the suborders was diversified into several lines that flourished in nearly all environments. Some produced formidable predators, others

---

*Let any who may doubt this examine the catalogue proposed by the paleontologists for our understanding of the various types of giant dinosaurs with respect their ecology and ethology. The latest, by Newman (1968), concerns the carnivorous dinosaur *Tyrannosaurus* which, we are to believe, did not stand erect, but at an angle to the ground, with the rigid part of its vertebral column horizontal and with the tail swinging clear of the ground, not constituting the third leg of a tripod supporting the animal's weight. It apparently used its very attenuated forelimbs to raise its weight off the ground. The tracks left by this enormous beast indicate short steps and a sinuous progress like a bird's: the tyrannosaur lived on carrion, and hopped along. So much for him—until the next interpretation.

herbivores, the biggest of all terrestrial creatures. We know of two-legged runners, *Struthiomimus* or lizard-ostrich-dinosaur, graviportal forms, bipeds such as *Iguanodon*, which grazed on the leaves of trees.

The hypertely of the dinosaurs and their total extinction within a fairly short time during the Danian is attributed to their gigantism. Nothing could be less certain, for not all of them were giants; *Protoceratops* of the upper Cretaceous in China was about 2.4 m long; the skulls of *Velociraptor* and *Saurornithoides* were less than 20 cm long, and *Anastosaurus* had a skull 40 cm in length, about the size commonly reached by present-day crocodiles.

While paleontology cannot identify the cause or causes of the extinction of such powerful animals, distributed world-wide, it does provide enough evidence to suggest that by their extreme characteristics the dinosaurs must have had great difficulties in adapting to altered circumstances and in finding sufficient food.

But their lines progressed into more and more hypertelic types for tens of millions of years. Such perseverance to attain an increasingly unbalanced state is a puzzling phenomenon, to say the least. If it was due to selection, we are therefore forced to admit that selection may occur in opposition to the interests of the species and doom it to extinction. Selection may be of the least fit, but that is not what Darwinism teaches.

Since the story of the dinosaurs was a long one, Darwinian selection for the greatest benefit to the species had plenty of time to channel the various lineages into nondefective lines. Why did it not do so? The chief reason might be that there were none. While it is unsure that the disappearance of these reptiles was due to their hypertelic characteristics, we do know that giant species of low fertility are frail and defenceless against microorganisms, inclemency of the weather, or climatic variations.

Elaborate adaptation makes the plant or animal highly dependent upon its environment and mode of life. If specialization becomes too narrow, it stops the living creature from altering its biotope or habits; it halts evolution and condemns the species to stagnation. It makes individuals increasingly less plastic and, in the event of a change of environment, jeopardizes their survival. The Darwinians say adaptation is the fruit of selection; if this be true, the inference is that selection very often does not operate for the good of the species but, on the contrary, its finality is vitiated.

Very cosmopolitan animals, such as cockroaches, rats, and man, are not subject to undue specialization. Their possibilities of survival in widely different environments insure for the species a considerable life expectancy.

INDIFFERENT OR USELESS ORGANS

Although, in general, we can fairly easily determine the role played by a given anatomical or physiological characteristic, or a given organ, the *uselessness* of some of them is quite apparent; they are mere details, decorations on the underlying structures. Examples of this are the pattern of leaf laminae, the color of the iris in the human eye, the consistency of hair (straight, wavy, woolly), and blood group, which have no bearing on the longevity of the individual or his adaptability to circumstance.

There has been controversy among orthodox Darwinians as to whether the black or brown stripes adorning the shell of woodland and garden snails (*Cepaea nemoralis* and *C. hortensis*) do or do not give rise to selection (Lamotte, 1951, 1959, 1966; Cain and Currey, 1968, Cain *et al.*, 1950). According to the color or pattern of the substratum, these stripes are thought to make the snails more or less visible to malacophagous diurnal birds. Quite recently in Perigord we observed the flat stones used by thrushes (*Turdus philomeles*) as anvils for cracking the shells of the snails on which they feed. The shells were *Helix aspersa* and *Cepaea nemoralis*, a uniformly yellow variety and a striped one (ratio 3:1). The shells collected were in the following proportions: 4 for the first, 1 for the second and 0.33 for the third. Those of *H. aspersa* blended much better into the background of stones and dry grasses than those of *C. nemoralis*, the commonest species. The thrushes hunted in the early dawn, when the sun was up and the dew abundant. Note that, except after rainstorms, snails chiefly come out at night, their greatest enemy being the hedgehog, which cannot see their colors in the dark. After feeding, most snails take shelter again where they are out of sight. It is to be noted that the advantage or the disadvantages caused by the stripes must be exceedingly small or even nonexistent.

Black-striped *Cepaea* shells, fossils from the Neolithic, have been dug up in the very same places where their descendants still live. The proportion of striped individuals there apparently exceeds that of the existing population, either because of a change of climate or because of genetic drift.

We should not exclude the possibility of the predator being conditioned to seek one kind of prey rather than another. In doing so it exercises a choice in which the lower visibility of certain prey is not always operative, as in the well-known case of the salmonid fish, the trout or char, in which some individuals of the same species feed on crustaceans (*Gammarus, Asella*) and others on insect larvae.

It matters little which of these explanations is the correct one. The effect of the presence or absence of stripes on the evolution of snails is certain, so the point is no longer of interest.

The shells of the gastropod mollusk family Cypraeidae are brilliantly and diversely colored, and porcelain-like. In the living animal the lobes of the mantle cover them practically entirely and conceal them. How could selection be at work on a hidden color? The pigment contains porphyrin and is possibly a product of excretion. Maybe selection is based on the chemical nature of the pigment and not the color. Hypothesis is added to hypothesis.

What differences does the shape of the nose make to the functions of smell and breathing? It is quite immaterial, and a European with an aquiline nose, or a snubnosed Papuan or a pygmy breathe equally well. And it is not a question of pleiotrophy;* the shape of the nose is in itself a hereditary, racial, and individual characteristic. Whether the iris is brown or blue, the retina functions just as well, etc. The same can be said of all racial characteristics (ear lobe, hair color, blood group, or whatever) and all individual ones as well.

Darwinian doctrine maintains that the bearers of these characteristics, so unimportant that selection ignores them, all participate in reproduction; the union between the progenitors is left to chance.

This is Weismann's (1886, see 1892, p. 399) *panmixia* "consisting essentially of the fact that not only the individuals in which the organ in question has reached the height of perfection, but all, regardless of whether it is properly constituted, are seen to achieve reproduction." This is assumed to be the roundabout way in which a useless organ regresses. Let us retain the only lesson to be derived from it, that the constancy of the insignificant characteristic testifies that it has no part in selection.

It has been objected that a characteristic trivial in itself may acquire importance for the individual or the species if linked to a gene which determines, at the same time, another useful or harmful characteristics. It is very tempting, but seldom rewarding, to invoke pleiotropy.

These minor characteristics are not linked to any state of inferiority or superiority, corporeal or mental.† In a general way the infinite variety of human genotypes due to differing combinations of racial and family characteristics cannot have any evolutionary impact, because it has practically no effect on the physical and moral health of the subject, nor

---

*Multiple phenotypic effect of a gene.

†It might be argued that in man they involve selection by virtue of the esthetic value (positive or negative) attached to them. This seems highly unlikely to be so; there is no decline in the variety of nose shapes in our societies.

really, *any utility whatsoever*. It is observed in the skeletal remains of fossil hominids and has contributed nothing of importance to anthropogenesis.

These are not vain arguments since they remind us that the variations represented by the alleles of a gene, modulations of the same anatomical or functional theme have no effect on constructive or innovative evolution.

UNFAVORABLE CHARACTERISTICS AND HYPERTELIES

Critically scrutinized, every living being is a mosaic of characteristics, some of which are essential, others merely useful, several neutral or indifferent, and a few harmful. There are many of the latter type in the human body: the extraabdominal situation of the testicles making them vulnerable to blows; the periurethral location of the prostate at the bottom of the bladder, causing difficulty in micturition when hypertrophied; the lack of resistance to crushing of the fifth lumbar vertebra; the weak-walled veins of the leg; the inability to regenerate tissues (with the exception of the intestinal mucosa, Malpighian layer, liver tissue); the nonrenewal and steady loss of neurons; the settling of cholesterol in the walls of the arteries, etc. The etiology of our noninfectious ailments exposes not only individual but species-wide imperfections in our organism.

A great many insects have inordinately developed organs that do more harm than good. There are *hypertelies* in most orders. The tendency toward an exaggerated pronotum (roof of the first segment of the thorax) is, in varying degrees, present in membracid homopteran insects; in some small strains it becomes enormous, bigger than the whole body, covering the entire back of the animal and bearing weird appendages: ramifying antlers, swollen in places, sloping forward or curving back. This overload unbalances the body and inhibits flight, which for these insects is short and zigzaged. Whether the pronotum is little or much hypertrophied, the species share the same habitat. Whatever shape it is, the organ fulfills its functions as a wall.

Hypertely of the pronotum exists in other homoptera, the cercopids, and other acridoid Orthoptera. In the latter, among the tetrigids it extends back to the tip of the abdomen, baring horns or raised like a knife edge high above the body! The same oddities are observed in other eumasticid acridians and, to a less marked extent, a few pemphagids.

Many scarabaeid coleopteran genera have the pronotum and head adorned with long and powerful horns, e.g., *Odontaeus, Bolbelasmus,*

Typhaeus, Copris, Chironitis, Onthophagus, Oryctes, Dynastes, etc. The hypertely in some cases affects the front legs. The large Indo-Chinese scarab beetle *Cheironotus* has forelegs twice as long as its body. They are even longer and thinner in *Acrocinus longimanus,* a long-horned coleopteran of Surinam. The hindlegs of the brenthid coleopteran *Calodromus mellyi* of Tonkin are huge, and those of *Anisoscelis foliaceus,* a coreid bug of Amazonia, are long and foliaceous. Other extravagances nearly as excessive exist in several fulgorids, tintigids, reduviids, and other hemipteroids.

The transformation of the rear wings into long straps projecting far backward beyond the tip of the abdomen has occurred in the flat-winged nemopterid Neuroptera (*Nemoptera, Nemopistha, Josandreva*); the wing is eventually drawn out in a long trailing thread, or a delicate hair in the genus *Nina.* Several butterflies show the same tendency (*Endemone, Syrmatis,* various *Papilio* spp.); their lower wings are narrowed and prolonged far beyond the posterior end of the body.

In various Coleoptera, hypertely affects the general body shape. The tendency for it to become very elongated and narrow is found in most brenthids, which are really like cylindrical sticks. In the case of a Brazilian mite, *Ludovix denticollis,* the rostrum, on top of which is a tiny mouth, is filiform and longer than the rest of the body. A surrealist vision!

The tipulid Diptera are characterized by long and fragile legs; some species are unable to walk, and their limbs merely allow them to hang from leaves and blades of grass. Rabaud (1932) shortened the legs of *Tipula oleracea* and found they were then able to walk. The evolutionary tendency toward longer legs does nothing to help the individual. Rabaud's study of cases like *Tipula* led him to confuse finality and utility. Had he referred to paleontology he would soon have realized that the two concepts do not coincide. An end may be useful, useless, or indifferent. The evolutionary trend may be good, bad, or neutral for the succeeding line. Evolution is indifferent to the survival or demise of the species. Faulty evolution is littered with the bones of its victims. The coexistence of a sound adaptation and a bad one, of useful, indifferent, or harmful characteristics, is one of the facts that demonstrate that the last word does not lie with selection in the world of living things.

Many groups other than those just mentioned have slid into hypertelic evolution. Among mammals, certain archaeocete Cetacea such as *Basilosaurus cetoides* (upper Eocene, Alabama) had a body that was exaggeratedly elongated behind the thorax to the point of giving it a snakelike appearance. Because of its structure, its vertebral column was extremely flexible. This extraordinary cetacean, dubbed "The Great Sea

Serpent" when first discovered, would appear to be the end of a line of which only a few links are known.

The camel-giraffe or alticamelid of the North American Miocene is a highly hypertelic line: small head, disproportionately long neck (the longest cervical vertebrae known), very long and slender legs, the hind ones as long as the forelegs. In all likelihood they were camelids, adapted to feed on the trees of the savanna lands, a sort of caricature of the giraffoid gazelles (*Litocranius walleri*) that now inhabit the dry savannas of Somaliland and Kilimanjaro and feed on the lower foliage of the mimosa tree.

Among the armor-plated edentates, the huge, squat glyptodonts, with the rear part of the body much higher than the forepart, covered in a curved, almost globular carapace formed by a mosaic of polygonal plates of bone, looked like giant tortoises or dinosaurs of the upper Cretaceous, *Scolosaurus*. These edentates, measuring up to 4 m in length (*Doedicurus*), lived in South America up to the Pleistocene and were known to Pampas man.

The pilous xenarthrans gave birth to a very specialized line that lapsed into gigantism: *Megalonychoides*, the ground sloths. The *Megatherium* of the Argentine Pampas reached 7 m in length and could stand on its hindlegs and browse on trees clutched by the forelegs, ending in talon-like digits of which II, III, and IV were extremely powerful.

But we should not lose sight of the fact that hypertely is by no means always unfavorable; the fact that many extremely hypertelic species, with a long past behind them, still persist, is enough to prove this. For these and many others, deformity does not upset the metabolic balance and consequently is not so important as might be supposed.

EQUILIBRIUM OF ORGANISM WITH ENVIRONMENT AND ADAPTATION

The semidominant and sex-linked *bar* eye mutation of *Drosophila melanogaster* does not endanger the life of the individual affected by it, but in a mixed population of bar-eyed and normal-eyed (so called wild-type) *Drosophila* in the same cage, it is found that "bar" females produce only 70 adult progeny whereas the wild ones produce 100. Geneticists express this by saying that the selective value of the bar mutant compared with the wild type is 0.70.

Although bar-eyed mutants have fewer descendants, this is not solely attributable to the reduction of their ommatidia, but to the fact that their mutants are weaker and less fertile, selection acts not upon organic loss or shape but on the accompanying *physiological deficiency*. This indicates that the *bar* mutation has no selective value in itself.

Selection maintains or restores the balance between population and environment, regardless of whether individuals are crippled, knock-kneed, or hunchbacked, just so long as they are sturdy and fertile. All else is purely secondary. Proof of this is shown by an insect, the planipennian neuropteran *Nophis teilhardi,* fairly common in the Sahara, with a long abdomen having a very sharply angled twist in one of the segments as if someone had tried to break it. Although this anomaly is pronounced, it has no physiological effect and does not endanger the future of the species.

This *real* aspect of selection differs from the one considered by geneticists who do not sufficiently distinguish between morphological deviations and the physiological ones not involving any change in the balance between organism and environment, or other equilibria—so much so that the numerous racial characteristics and their individual modulations in man persist and do not appear subject to selection (we will ignore populations that undergo heavy natural selection, such as pygmies, American Indians, and Australian aborigines).

It is better to have a powerful immune system than the muscles of Hercules or the anatomy of Adonis. Moreover, selection eliminates the bearers of characteristics that diminish fertility, undermining resistance to aggression from climate, disease, parasites, and so on.

Details of forms and functional mechanisms are unimportant to the development of the species. The truth of this proposition is demonstrated by the continued existence of systems, physiological and behavioral peculiarities, which we consider to be aberrant indeed. Rabaud (1922) certainly realized this, but with extreme exaggeration denied the reality of adaptation, which is absurd. However, what survives from his work is the notion of the *persistance of the noninjurious,* which is important to man, animals, and plants, together with the idea that what is important for the living being is to have a positive physiological balance sheet.

Given the low utility of many coadaptations, it is hard to see how selection could have presided unaided over such delicate fittings as the double lock of the ant lion's mouth, the hinges of the bee's wings, the push button of *Nepes*, etc., which have to be perfect if they are to function at all.

For a population to subsist it must be stable or increase. What is important for the evolutionist is to know the causative factors in the demographic balance sheet. A simple matter in theory, it is difficult in practice because we always have great difficulty in identifying the environmental parameters and genes which cause the individual's inferiority or superiority with respect to his kin.

In the light of recent research we note that the number of neutral characteristics, i.e., those of no selective value, is higher than was thought. We have cited some of them; they were innumerable. Although of no importance in evolution, they are nonetheless constant and species-specific. Specific characteristics are likely to be most neutral.

Genic (or intragenic) diversity differentiates individuals, and confers upon each its own originality. Certainly many parts of the individual vary, but not to such an extent as to give rise to selection. The mutations which only slightly upset the sequence of bases in the DNA segment we call the "gene," and consequently modify the proteins it encodes, pass unperceived; they have no evolutionary value (the neutral mutations of the Darwinians).

VARIATIONS CAUSED BY THE ENVIRONMENT AND THE ACTION OF THE GENES ACCORDING TO THE ENVIRONMENT

Agriculture and stock breeding place huge populations at our disposal and thereby enable us to measure mutability, and the transition from one environment to another supplies a means of revealing the living creature's ability to adapt to circumstances, of evaluating its self-regulating capability and the expressivity of genes in respect to environment. The environment sometimes has effects of considerable amplitude on plants. For example, cultivation of plains flora in mountain areas revealed to Bonnier (1890, 1894a,b, 1898) facts that are classic and highly interesting. Some are unaltered in appearance whether grown in the Paris region or 2000 m up the slopes of le Lautaret. But the corniculate bird's-foot trefoil (*Lotus corniculatus*) and the self-heal (*Brunella vulgaris*) are completely metamorphosed and assume an Alpine facies while shrinking in size by internodal shortening. Others, the Jerusalem artichoke (*Helianthus tuberosus*), for example, adopt, instead of the upright position, the rosette style so often affected by Alpine plants. At high altitudes, the leaves of lowland plants grow smaller but thicker and richer in chlorophyll; pigment formation is stimulated, hence the brightly colored blooms. There is a considerable extension of root systems. Yet plants of a third category cannot stand transplantation to another habitat, and die.

There are plenty of other examples of the transformation of a plant under the influence of the environment. *Veronica anagallis*, normally a subaerial air plant, alters in appearance and metabolism when grown under water. The adventitious twigs and roots grow larger, and leaf surface and hairiness of the stems increase. The bark thickens, the central cylinder dwindles, and stomata are smaller and fewer. The two epidermises in the leaves are charged with chlorophyll and there is no palisade tissue; woody growths and cuticles are reduced in size. Water and mineral content rise. The nitrogenous metabolism is not the same when the plant is grown in the air or in water, and the plant's absorption of nitrogen and phosphorus and its synthesizing ability are increased. Underwater photosynthesis is twice as great as in air, and lipogenesis also increases (Gertrude, 1937).

One might argue that these characteristics are not hereditary; this is both true and false. This is not necessarily a contradictory statement; it must be interpreted from the facts as a whole. It can be explained as follows: Plants subjected to these experiments have the same genes as the controls. *Some genes work differently according to the environment in which the individual is placed:* the characteristics upright habit, thin leaves, etc., assumed by the Jerusalem artichoke on the lowlands are no more hereditary than the rosette habit and thick leaves it assumes in mountain areas.

Autoadaptative variations, which are multiple, prove that numerous genes have effects that vary according to the environment in which the organism is immersed. The experiments by Bonnier and his disciples prove this variability.

The chain of chemical reactions set off by the genic enzyme differs according to the values of climatic and other parameters (temperature, humidity, light intensity, circadian rhythm, and so on) which no doubt modify the cytoplasmic environment in which the genes operate. Experience forces us, therefore, to concede that the rule that a gene always determines the same characteristic unless it mutates is over-rigid.

Studies of the action of light intensity on plants, longer and shorter days, alterations of hot and cold under varying hygrometric readings, as well as countless experiments in phytotrons, those sophisticated greenhouses in which the scientist can imitate climate at will, not only reveal new physiological facts (flowering and fruiting, tuberization, plant cycle, etc.) but also the adaptive capacity of the species studied, i.e., the range of reactions by certain genes to counter the effect of the environment, whether or not there is aggression.

It is likely that the gene conveys the same information in each case but the responding materials react differently according to the circumstances. We cannot be sure that messenger RNA always finds the same amino acids, ions, and other appropriate substances for the genesis of specific proteins; the environmental conditions modify the cytoplasm, the effector: *The responses to messages are not invariable.* In this area there is much, or indeed everything, to learn.

Thus, the above-mentioned experiments on plants show us that (1) certain specific genotypes (classically these amount to DNA molecules) produce effects variable according to circumstances; (2) only certain metabolisms triggered by the genes are modified; (3) certain genes by their effects show great stability, whereas others are highly variable.* The same gene does not have the same effect in all individuals; this is what geneticists call expressivity. They also note that a gene does not manifest itself in all the individuals bearing it, as Mendel demonstrated in his epoch-making experiments.† The term *penetrance* denotes the frequency of cases in which the gene has an observable effect.

The "lobe-eyed" mutation of *Drosophila* is dominant in heterozygotic individuals but only crops up in 75% of its bearers, which gives it a penetrance of 75%. Expressivity ranges from the normal eye to none at all, with all the intermediate states, including the deeply notched eye; the gene's expressivity is thought to depend either on the composition of the genome or on the cytoplasmic composition of the effector cells involved in building the eye.

A gene has a fixed index of selectivity, provided it produces the same effect in a stable environment. The case of variable effect genes is different; the coefficient of selectivity varies with its expression (cf. lobe-eyed *Drosophila*). Most Darwinians acknowledge that the characteristic expressed remains constant, whatever the environment. This is an oversimplification, which overlooks vital facts and detracts from the value of the calculations made and the conclusions proposed.

The above facts show that certain genes determine different characteristics that are generally adaptive according to circumstances. This means that they have their own regulating capability and are possibly included in operons or similar systems.

Are there any responses of the organism not directed by the genes? Certainly, there are. Most physical or chemical stimuli affecting organic functions—sensation, food, and so forth—do not get as far as the gene.

---

*Not all genes would appear to be equally variable; consequently, evaluation of possible variants according to number of nucleotides and the bases they contain is probably of theoretical significance only. In this area prudence should be the rule—since some changes are immediately incompatible with life, the ovum may die shortly after it is fertilized, and it may be hard to observe.

†Dominance, with its pluses and minuses, results from the reaction between genes of the same allelomorph pair.

But they do modify the cellular environment and thereby the effect of the genes.

The idea that the germ is independent of the soma is an exaggeration. The gene does distribute information, but not just any time or in just any environment. The chemical informers, the hormones, stimulate or inhibit the gene, *for the endocrine glands convey information of their own,* although as parts of the organism they are dependent on the genome (cf. also Chapters VIII and IX).

The genes convey primary messages, and some organs built to their order are themselves able to generate information and send messages to the other parts of the organism. This is the case for the endocrine glands, as we have just pointed out, and during embryogenesis for the organizers (primary, secondary, tertiary). The pheromones secreted by the tegumentary glands create an aura of information around the organism that seems to increase as our knowledge of animal ethology and physiology grows greater.

Thus, in multicellular organisms the genes relay their information in cascade to particular cells or organs, which become second-order informers and initiators of synthesis. The hormones and other substances act in turn on the chromosomes and in this way stimulate the activity of certain genes. Practically nothing is known about the mechanism of their action. Inside the cell actions and reactions occur incessantly, and amid such activity DNA cannot be said to be all-dominant. If its informer system were insufficiently flexible, the organism living in an environment of unsteady parameters would be doomed to almost immediate death.

According to Darwinism, which has adopted the dualistic concept of the body of multicellular organisms as enunciated by Weismann (1885, 1888), the reaction involves only the perishable part of the individual, the soma; the germ is inaccessible to alien influence except for the mutagenic agents; hence the notion that the germ is passive in evolution, registering the random blows which are struck against it. The bodily reaction leaves it unchanged. Its only possible, and passive, counterattack is that it contains mutated and randomly preadapted genes. Agreement between genes and environment is a matter of pure chance. The living creature submits to its fate. Like a rudderless boat, it floats on the ocean of time and docilely and indefinitely obeys the aleatory movements of the waters supporting it. Such is the Darwinian concept.

In my opinion, it does not correspond to reality. The genesis of control systems with no intervention by the organism is difficult to imagine and postulates Himalayas of chance occurrences that, moreover, could not have happened for lack of time and a sufficient number of generations. Take, for example, regulation of the coagulation of blood, a highly complex phenomenon to which biologists seem to have given little thought.* Its normal cause is the opening of a vein, artery, or capillaries; the blood brought into contact with the lip of the wound (damaged tissues) becomes the site of chain reactions ending in the formation of a clot. This is only possible because there *preexist* in the blood reaction agents or their precursors whose end effect is to coagulate certain proteins of the blood plasma. The organism, ready for all eventualities, bears within itself *in the latent state* its own protective system. Genes control the elaboration of coagulants, proteins, and enzymes.

Such a process forms a single whole; a lack of a substance arises, an enzyme is affected, and the system will not work. One does not see how it can have been formed by successive chance effects supplying a protein or an enzyme in any random order. Besides, we know that the effects of mutations on the system are disastrous and form the lengthiest chapter in blood pathology.

The system has become *functional only when all its components* have come together and adjusted themselves to one another. The Darwinian hypothesis compels us to postulate a preparatory period during which selection acts upon something that does not, physiologically speaking, yet exist. Under the necessary conditions of the postulate, the action can only have been prophetic! Any explanation ruling out the active intervention of the organism in the acquisition of regulating systems may be regarded as inadequate.

To take as an evident truth the fact that the control mechanisms attenuating or neutralizing the actions of the environment (these are, let it be remembered, complex systems having several coordinated elements) could have been assembled by successive and lucky strokes of chance without the slightest need for the organism to play any role whatsoever, is to sacrifice objective scientific analysis to a wholly verbal magic trick.

PREADAPTATION AND SELECTION IN CLOSED ENVIRONMENTS

Darwinism attributes increasing significance to preadaptation, whose dual chance meeting of a mutation and the ability to meet a need suits it perfectly.

*An even more demonstrative model would be immune mechanisms, but these are so complicated that even an abridged description would have strained an already heavily burdened text.

New life was infused into the idea of preadaptation, notably by the demonstrations by Luria and Delbrück (1943) and Lederberg and Lederberg (1952) of the preexistence in bacterial populations of individuals on which a chance mutation has providentially conferred resistance to a virus (bacteriophage) or some drug synthesized by man or extracted from a plant. The preexistence in a population of mosquitoes or flies of individuals resistant to DDT and other synthetic pesticides corresponds with respect to metazoans to what has been observed among bacteria.*

The notion of preadaptation has been extended to the animal and plant kingdoms as a whole. The virtual possibility of a population going from one environment to another, to ward off aggressions, was regarded as solely due to the presence of individuals *preadapted to live in the new environment or put up with changes in the old.*

The quasicertainty that the resistance of bacterial populations to antibiotics or viruses originates in preadapted mutants is a thing of the past as far as an indisputable and complex evolution outside our time is concerned.

So as not to lose ourselves in theoretical speculations, let us take a look at some concrete cases.

Cavernicolous Insects

The fauna found in caves and fissures in the earth is alleged to comprise preadapted strains. For example, nonaquatic cave-dwelling insects that need a moist and still atmosphere, low and quasi constant temperatures, and whose perceptual universe is based on the reception of chemical (especially olfactory), mechanical, and hydrous signals, formerly lived on the ground under the shelter of mosses and dead leaves. They

---

*The presence in a population of bacteria of individuals preadapted to resist viruses, or destroy a drug they have never come into contact with before, ought seldom to occur. This is, indeed, true in most cases. But there are disconcerting exceptions. Before 1958 most strains of *Gonococcus* were inhibited by penicillin administered in a dose of 0.2 units per cm^3 of culture. Today it takes at least 0.5 units. This tendency to become resistant is true of other biotics. It was believed that tetracyclines would have a more lasting curative effect. Not so; resistance to these drugs spread even more quickly than in the case of penicillin. In 1965 concentrations of less than 0.5 μg per ml inhibited 77% of the cultures tested; only 1% required concentrations of over 1 μg/ml. But by 1970, 0.5 μg/ml would inhibit only 16% of the strains present, and 27% required over 1 μg/ml. One strain studied in 1970 went on growing after exposure not only to concentrations of tetracycline of 3 μg/ml but also to doxycycline, a compound similar to tetracycline and widely used (cf. Sencer, 1971). The same thing has happened in the case of sulfonamides, which 25 years ago worked with gonococci. The ease with which these different drugs are resisted is astonishing. Is it only a matter of preadapted individuals? Maybe so, but one would like to be sure.

were *preadapted* to living in caves, into which they stumbled in the course of their random wanderings. Blind or microphthalmic, they found themselves in an environment that suited their infirmity, which was canceled out by the darkness, and no longer had to fear competition from better-endowed species. Most families, subfamilies, and tribes of cavernicolous insects number among their ranks both hypogean species and those living in the open air. The subfamily Bathysciinae, catopid Coleoptera of the superfamily Staphylinoidea, is a good illustration. The 700-odd species or subspecies that compose it are in some cases epigean, in others endogean (living under stones and in the soil), or hypogean, living in caves and cracks in the rocks.

The argument is that the forest mosses and humus were for insects the antechamber to caves. Some Bathysciinae still live there, whereas others have migrated into the depths of the earth; their microphthalmia, the loss of one larval ecdysis (3 instead of 4 as in all the other staphylinoids), and their hygrophily predispose them to subterranean living. A few species form a transition from true epigean forms to troglobionts. One of these *Bathysciola schioedtei* lives in mosses or in the entrance to caves.

Cave dwellers are characterized by the loss of their eyes (the micropthalmia of epigean species is thus considered a first step in the process), by compound sensory bristles, by the length and slenderness of the limbs, the formation of large eggs, the shorter duration of postembryonic development, and longer imaginal lifetime. Losses and gains more or less cancel out according to the zoological group considered.

It seems certain that the Bathysciinae continued their evolution underground. Their modes of reproduction prove it. *Speonomus delarouzei* of the eastern Pyrenees lays average-sized eggs, the first stage larvae live and feed freely; after ecdysis in a small cell which it builds itself, it continues to wander about but takes no food. The female *S. longicornis* lays, in a rhythmic cycle, a single large egg covered with agglomerated chalk particles, and the offspring turns into a nymph in its shelter without venturing abroad or eating. Between these two species are classed the bathysciinines that lay only one egg but have a larva which, without feeding, leads a wandering life before building the little cell for nymphosis, as does *Leptodirus hohenwarti* which lives in the karst caves in the northern Dinaric Alps. The larvae of species with a shortened life cycle have an atrophied nonfunctional digestive tube. They undergo an "anatomical simplification" [to use the term employed by Sylvie Glaçon (1953), who discovered their cycle in 1953]; their sensory bristles become simple instead of compound, their mandibles symmetrical instead of asymmetrical, they lose their paraglossa and retinaculum, and their cercus has only one joint. Between the epigean forms and the Bathysciinae most thoroughly adapted to their subterranean habitat are intermediate species which show that the evolution of this short line *took place gradually in the caves.*

The imagos, less affected than the larvae except for loss of pigmentation (already visible in epigean species) and their blindness, are not very different from lucicolous species. The greater longevity (*Speonomus* lives 4–5 years) and loss of nycthemeral rhythm recur in nearly all cave-dwelling insects. Nevertheless, the cavernicolous species is less dependent on cave ecological conditions than was thought: it is rather surprising to learn (Deleurance-Glaçon, 1963) that Bathysciinae are so little disturbed by light that they can be bred in daylight and at temperatures of hardly less than 16°C. Their greatest dependence is on atmospheres of high hygrometric values. This relative adaptability notwithstanding, we are practically sure that Bathysciinae and all other true troglobionts are incapable of living outside a subterranean habitat.

It would not be easy to find a Darwinian explanation of this evolution. What "necessity" is there for a cave-dwelling insect to increase its specialization, which makes it a slave to its environment, when it can live quite well inside the cave in its preadapted form?

Whatever the explanation, the results of all the "experiments" conducted by the cavernicolous species are satisfactory because they go on living there, and no one has any reason to think them degenerate. Their specialized physiology condemns them to a restricted environment, a feature they share with many other species. Do we call the koala degenerate because it finds only the leaves of four species of *Eucalyptus* to its taste, or the blepharocerid larvae degenerate because they are strictly limited to stream water? By juggling value judgments we can argue the opposite view and think of cave dwellers, not as degenerate, but as so perfectly adapted to a special medium as to be unable to put up with any other.

More than one species of the same genus seldom inhabits the same cave. The *presumed* reasons are multiple; one is interspecific competition, a second the isolation of caves and potholes, yet a third, but more unlikely, the specific ecological conditions of the cave or territory considered. Species of different genera compete together in caves just as in the open air, but less fiercely.* But between species in the same genus competition is so severe that only one is left as victor.

Is it true that interspecific competition is accompanied by selection to aid greater adaptation to cave dwelling? We cannot be sure. Such puta-

*We are told that species of different genera exploit separate ecological niches, so that there is no competition.

tive selection applies to negative characteristics since the most "regres-sive" species are the most highly evolved. The notion of "useful" and "favorable" is highly relative here. An individual well-adapted to the environment may not be a good fighter and may give way to a stronger and more aggressive adversary less well-adapted than itself. The reality, when closely examined, is complex and the effects of the struggle for life on the evolution of cave-dwelling species highly relative.

The pselaphid coleopteran studied in minute detail by Jeannel (1952, 1953) is an instructive case. Noncavernicolous species either live in humus or under sunken stones (endogeous species). They have a marked sexual dimorphism; the females are apterous, microphthalmic or blind, the males are winged and have normal eyes so they move around away from their native milieu to look for females.

The cavernicolous pselaphids, such as *Glyphobythus*, retain the habits of the endogenous forms and the males leave the cave to breed. In a few cavernicolous pselaphids there are two kinds of male, one normally equipped with wings and eyes and the other microphthalmic and apter-ous (*Octozethinus leleupi*). In the humus of the Zairean forested uplands, relatively cool and damp, we find Pselaphidae at the same evolutionary stage as the circum-Mediterranean cave-dwellers. There are two possi-ble explanations (Jeannel and Leleup, 1952): (1) In Zaire, adaptation to underground living is just beginning whereas around the Mediterra-nean it is over. (2) The Mediterranean subterranean pselaphids may have undergone all their evolution in the forest humus covering the major part of Europe in the Tertiary era, and sought asylum in the caves when the climate turned dry and colder. In the absence of a fossil record we cannot explore the past but are obliged to frame hypotheses on flimsy and narrow bases.

What variations enabled the strains to change? One immediately thinks of mutations, especially since several characteristics of cave dwellers, like so many other mutations, are negative: loss of pigment, eyes, compound sensory bristles, etc.

But the variations in the cave-dwellers have not been nondescript, but have accentuated those prefigured in the epigean species. They have constantly followed the same directions, and have added up in the course of time (losses or gains of organs or functions). The existence of intermediates between the epigean species and the most evolved hypo-gean ones proves this. Moreover, entomologists have acknowledged that apparently natural lines keep to precisely delimited geographical areas.* So it is not at all a matter of disorderly, nondescript mutations

*The fauna of caves includes a very high number of endemic species of which it is not easy to say whether they are paleoendemics seeking refuge or neoendemics recently individualized in a modest-sized geographical area.

undergone for millions of years, with no evolutionary effect, by *Drosophila,* the mouse, and bacteria.

The abundance of microphthalmic or blind, depigmented species in subterranean species in general is a well-established fact: Turbellaria, Crustacea, arachnids, insects, fishes, urodelan amphibians, etc., all have many cave-dwelling species, microphthalmic or blind, as the case may be. Why wonder at so many blind varieties? say the Darwinians. Eyes are no use in the dark, so that not having any is no loss; they do not need to be selected. But then, if the blind and the seeing are at equal odds in the caves (cf. the hypothesis of panmixia of A. Weismann discussed earlier), why do the former prevail over the latter? That is the snag there is no getting around.

Let us say that blindness is also accompanied by depigmentation and lower metabolism. Why do most cavernicolous animals display all three characteristics? Selection of mutations, plus a bonus in their favor, is the answer, which takes little imagination to find; but *what it implies is that the characteristic triad has occurred in all groups of cave-dwellers.* Three different kinds of mutation, interlinked and occurring in all groups, even if not in every species, are rather a tall order even for chance, which has its limits, after all.

This evidence strongly suggests that some other model of cave-dwellers will have to be proposed, without all these fanciful trimmings and based firmly on observed and unchallengeable facts.

Cavernicolous Fishes

Cave-dwelling fishes do shed some light on evolution, though less than the hypogean insects. Comparatively few in number and distributed among nine families, four of which belong to the siluriforms, their species are geographically isolated and mutually independent. Some have been found at only one site. Part of our difficulty is also that very little is known about underground watercourses. Among cave-dwelling fishes having a somewhat more extensive distribution, three genera are mentioned: *Amblyopsis* (Kentucky, Indiana, Missouri), *Typhlichthys* (Oklahoma, Missouri, Arkansas, Kentucky, Tennessee, Alabama), and *Caecobarbus* (lower Zaire).

The habitual segregation of cave dwellers because of the separation of

---

The species of *Speonomus,* a genus found only on either side of the Pyrenees and in Sardinia, has small distributions: *Speonomus delarouzei* has only been found in caves in the eastern Pyrenees, *S. longicornis, S. pyranaeus,* and *S. abeillei* in Ariege, *S. alexinae* on the Atlantic slopes of the Pyrenees, and so on. This is true of most genera of Bathysciinae.

hydrographic systems underground has very likely, and to an unknowable extent, influenced the transformations they have undergone. The fact of their segregation means that evolution has only affected spatially highly limited populations. We cannot trace their time scale nor recognize their phyletic descendence. We know there are epigean and hypogean, species, but the intermediate steps escape us.

This is an important observation, which reminds us that in the very numerous interpretations proposed for the origin of troglobiont fishes, imagination plays the biggest part. It is hard to account for the restricted occurrence of these fishes. Why should barbels (*Barbus sensu lato*) have no cave-dwelling species in Europe, but two genera *Caecobarbus* (lower Zaire) and *Barbopsis* (Somalia)? The European barbels (*Barbus fluviatilis, B. meridionalis*) have often been carried by flood into underground river systems but, for reasons unknown to us, possibly ecological, have not established strains in them.

The evolution of fishes in underground environments has been almost entirely reductive: loss of eyes and of tegumentary pigment and a smaller brain, with one single gain, greater development of the tactile and vibratory senses. The term degeneration has been applied to it, but such a pejorative description hardly suits healthy and vigorous animals. It is primarily a case of very narrow adaptation to a particular environment.

Weismann (1886), as a faithful follower of Darwin, posited an ingenious interpretation of this evolution, taking as his example *Proteus*, a perennibranchiate amphibian, and a cave fish of Kentucky, both of them blind. He surmises that *Proteus* derives from an epigean newt, normally oculate, that penetrated underground and found itself "able to find food without using its eyes" (we have here already an application of the principle of "physiological" preadaptation). The regression of the eyes is not due to failure to use them; as selection between individuals with better and poorer sight is stopped, each has as many chances as the other of persisting and reproducing. Subsequent cross-breeding will take place, and the result will be a general weakening of vision. "As soon as natural selection no longer maintains the eye at its highest degree of organization, it is *necessarily* doomed to become decadent" (my italics) (Weismann, 1886, see 1892, pp. 395–396).

Hence the involution of the organ is said to be due to the lack of selection; all individuals in the population, regardless of the state of the particular organ, take part in reproduction.* Weismann calls this *pan-*

---

*It is hard to be sure that in any given species copulation is random in character. For example, female *Drosophila*, formerly thought to indulge in total panmixia, on the contrary select the male by specific stimuli.

mixia. The theory has not been discredited in Darwinian circles. It has been pointed out that its consequence ought to be to arrest evolution, but even in a panmictic population, a few mutations are sufficient for its continuance.

Panmixia does not *ipso facto* involve the disappearance of the selectively neutral organ, even when this becomes vestigial; e.g., the cecal appendix, the muscles of the outer ear, and the coccyx remain unchanged. Mutation causes the immediate and total disappearance of the organ; evolution acts slowly, and durable vestigial traces remain.

The tapeworm, which has no digestive, circulatory, or respiratory systems, is not a monster nor a degenerate, nor is *Sacculina,* which goes even further in its organic transformations. Any strict adaptation to a mode of living or a particular environment carries with it a loss of organs not functionally essential to life. It is as if the living creature, economizing its store of energy, eliminates organs and functions that in existing circumstances do not aid survival or reproduction.

Although in the case of the cave fishes adaptation is mainly accompanied by losses, it is not the same for cave insects, which gain by it, e.g., greater development of the organs of touch and receptors of mechanical vibrations. The "parasitic degenerations" reported among the Platyhelminthes also involve indisputable gains: attachments, absorption of the food in which the parasite is bathed by tegumentary microphagocytosis (= pinocytosis), multiplication of the genital organs, enormously increased numbers of eggs, multiplication of larvae by paedogenesis, budding, and anaerobic fermentation of glycogen.

What is the adaptation of a cave dweller but the ability to survive in a rather cold environment, very wet and dark and poor in food resources? What necessity can there be of increasing its dependence on a constant temperature and humidity? None, so far as I can see. The animal becomes totally dependent on its environment, and in the event of the slightest change is liable to perish. Lower fertility (only a single egg in some cave species!) indicates low interspecific competition, so that the species is content with but a few gametes which, being larger, produce young whose organization is more advanced than those hatched from small eggs.

What role is left for selection to play if the animal is already adapted before it gets to the cave? None, apparently—all the offspring of preadapted animals taking to caves are from the start matched to their environment.

As for the convergence of characteristics in the different strains of cave dwellers, chance alone is unable to account for it. The orientation of troglobionts is so clearly preordained and faithfully followed that one wonders whether factors other than those so far considered may not be the true agents of their evolution. No wonder, then, that Vandel (1964), who knows more about subterranean environments than anyone else, should consider the history of cave species to be a manifestation of *orthogenesis,* or guided evolution, like so many others revealed by paleontology and biometrics (Matsakis, 1962). The environment being constant or almost so, the orthogenetic variation may be inferred as proceeding from the organism itself. Disorderly and inadequate, mutations are by their very nature opposed to guided evolution. Orthoselection, as imagined by Plate (1913), is impossible since it can only function provided that the mutations occur in a certain order, which they never do.

PREADAPTATION AND ITS ROLE

It is likely that predisposition played some part in populating the subterranean environment. Since penicillin, aureomycin, or DDT is absolutely not involved in the genesis of bacteria, mosquitos, or flies resistant to them, we need not go to caves for a model of troglobitic insects, crustaceans, or fishes. Preadaptation and chance take care of that.

By extrapolation from one case to another, logic concludes that there was no need for a cave in the genesis of cave species. Taking the generalization even further, we can accept eye formation without light and independent of any photoreceptor function. At a closer look, this explanation makes any invocation of antirandom selection unnecessary or superfluous, since chance alone suffices. And why should not the whale have been formed as a land animal, and the tapeworm outside the digestive tract of a vertebrate? There is no logical reason why not. We are very wary of such an *argumentum ad absurdum.*

Although a population threatened with destruction by a cataclysm may on occasion be saved by its randomly occurring preadapted mutants, there would seem no reason why such exceptional variants would generate a cotinuous guided evolution as observed in plant or animal lines. Who would dare argue that the marvelous adaptation to aquatic life and the plumbing of the ocean depths that we see in the whale represent a random collection of properties themselves aleatory, in harmony, always by chance, with an environment and mode of living not yet adopted by the animal? Yet another random event was the preadapted mammal's stumble into the water, where he liked it so much he decided to stay!

What audacity to claim that ticks, *Sacculina,* parasitic copepods, *Bopyrus,* and *Portunion* and other parasitic isopod Crustacea were formed, lock, stock and barrel, without ever encountering the host animal reserved to them by "Fate."

Such fairy tales, like those told by my grandmother for my amusement, are not to be taken seriously. This utility is to show the excesses into which we may be led by a doctrine not kept within bounds by reason or concrete observation. Preadaptation must reduce its pretensions, or lose all credibility.

The chance preadaptations noted so far are confined to a single property (elaboration of an enzyme by a mutated gene) and have nothing to do with a set of coordinated features; their evolutionary importance, if any, is thereby seriously reduced.

COAPTATION AND ADAPTATION OF UNLIKE PARTS

The humblest creature often poses evolutionary problems in stark terms that cannot be escaped by mere rhetoric. None is more "antichance" than the ant lion larva, for it offers the naturalist an exceedingly rich collection of coaptations and in all its organs pushes specialization, both morphological and physiological, to an extreme.

Its head is flat and its mouth has become a slit hermetically closed by a double lock, two astonishing coaptations of parts formed *independently* of one another during ontogeny. The forward edge of the head is bent back underneath to form the upper lip, while a hexagonal plate whose rear edge is welded to the wall of the skull forms the lower. The roof of the buccal slit has a triangular projection that fits into a preexisting hollow of the lower lip, the whole forming a kind of pushbutton. A second, complementary locking device is a dovetail assembly in which a sideways and forward projection of the floor of the pharynx is lodged in a corresponding groove in the roof of the mouth.

A closed mouth is admittedly not conducive to absorption of food, and rules out all solids, but the ant-lion larva has adopted a very special means of nourishment: It pumps out the body fluids of its victims. Because of the locking of the buccal orifice, the mandibles and maxillae are pushed laterally rearward from the oral cavity to the edges of the slot. They are highly elongated and their free extremities curve inward to form sharp-edged pincers. The mandibles have a longitudinal groove on their inner face running as far as the root of the appendix, thus communicating with the buccal cavity. The maxilla, reproducing on a smaller scale the curve of the mandible, runs parallel to its inner side; externally it has a gutter, which fits extremely snugly to the groove in the mandible. The superimposed trench and gutter form, edge to edge, a capillary channel: Coaptation is perfect.

The ant lion larva, that demon of the sands as the entomologist Wheeler calls it, digs its buccal pincers into the body of its prey. It first injects a paralyzing poison secreted by a gland and kept ready in a small basilar swelling of the maxilla; the poison travels not down the capillary channel but along a longitudinal gap between epidermis and cuticle; thus, it does not mingle with the digestive juices. Once the prey is paralyzed, the ant lion injects into it, via its maxillomandibular capillary tube, digestive juices that attack and liquefy viscera and muscles. The juices can be withdrawn and then the body juices of the prey sucked off by conversion of the pharynx to a reciprocating pump.

The larva's sensory equipment includes numerous organs capable of recording the tiniest mechanical vibration of the support. Let but a grain of sand trickle down the walls of its tunnel, and the larva, instantly alerted, gets ready to leap on the possible prey with its mandibles gaping wide.

To struggle with the prey, the hunter needs to get a firm grip on the substratum. It is able to do so because the rear end of its abdomen bears chitinous hooks that bite like teeth into the substratum and prevent slipping.

The ant lion's habitat does not support a plentiful fauna, and starvation often occurs. For days on end the sedentary larvae lie hidden in the sand at the bottom of their funnel-shaped trap, on the lookout for possible prey, ant or other insect. But their physiology enables them to fast for long periods without dying.

Since the larva lives in very dry sand, often sheltered from the rain, it can only survive by avoiding all losses of moisture. Its excretory function operates economically and resorbs the water containing its urine and other waste. What is more, the digestive tract ends at the junction of intermediate and lower intestines so that all defecation becomes impossible and the water absorbed with the food is totally conserved. In addition, the ant lion larva is protected against evaporation by exceedingly impermeable integuments.

Obliteration of the rectum from the digestive system enables it to be transformed into an organ having a new function, that of silk tank and spinner. It dilates into a blister whose tip forms a fold into which the free extremities of the excretory organs or Malpighian tubes (cryptonephridism) penetrate, an arrangement facilitating resorption of the water in the urine. The segments of the Malpighian tubes nearest to their insertion into the intestine alter in function, no longer excreting waste but secreting silk made of proteins and accumulated in the reservoir of the ex-rectum. Through the very narrow anus, which serves as a spinneret, the fully grown larva ejects the silk and, like a caterpillar, spins a cocoon in which, a few weeks later, it undergoes metamorphosis into a perfect imago.

So we now have to turn to the Darwinians and ask: "Have you ever seen a mutation simultaneously affecting two separate components of the body and producing structures that fit one another precisely? Tell us, have you ever beheld three, four, or five simultaneous mutations with matching structures producing coordinated effects? And yet you have observed and described thousands upon thousands of mutations. The huge populations of animals and humans bear witness to their frequency. In any man the number of mutated genes is extremely high. The mutations are nondescript, monstrous, or pathological, and are invariably, repeat, invariably incoherent. And yet it is by that that you claim to explain the biological order, and make evolution intelligible?" These are vital questions that demand an answer. There is no way of getting around them, or evading the issue. Every biologist who wants to know the truth must answer them, or be considered a sectarian and not a scientist. In science there is no "cause" to be defended, only truth to be discovered.

How many chance occurrences would it have taken to build this extraordinary creature that braves the burning sands of the Sahara, endures prolonged fasting, economizes water, detects the slightest vibration in the ground, lies in wait for days on end at the bottom of a funnel, or goes forth, freely, to hunt down its prey?

It is not enough for a property to appear, it has to come at the right time. These accidents, always fortunate of course, produced their effect by occurring in a certain order, for, out of order and untimely, they would have remained imperative. What scientist would venture to estimate the chances of such a cascade, such an avalanche, of coordinated and mutually adjusted chance occurrences? The odds are infinitesimal.

Please remember, too, that the case of the ant lion is not at all an exceptional one, chosen to support a thesis; such an accumulation of adaptations and coaptations is the rule. Consider the human skeleton and its ball-and-socket joints, which fit with perfect precision, and you have another example. The copulatory organs of the arthropods reach an amazing accuracy in their coaptation between the sexes, notably in the case of dragonflies (Odonata), araneids, and so on, adapted, as well, to behaviors as unexpected and illogical as they are diverse.

FINALITY AND ADAPTATION

All of life's phenomena are reduced in the final analysis to physics and chemistry. This much is crystal clear.

The vast majority of the substances that go into the making of living beings or enter into their functioning (supplies and transfers of energy, syntheses) contain the metalloid carbon whose special properties render it capable of an infinite number of combinations, attaining highly complex structures such as no other simple element whatsoever is able to achieve.

Living creatures, by their structure, architecture, chemical nature, and physical state, acquire properties differentiating them from inanimate bodies without *ever* going beyond their physicochemical framework. *The complexity of the structures and their components brings out the properties that are inherent in them and consequently not available to inanimate bodies.* Life is movement, the balancing and unbalancing of the parts of a whole, harmoniously joined, reacting upon one another or else upon environmental influences.

For all the innumerable studies made of them, the properties of the living creature are still not fully known. This is self-evident in the case of the nervous system, behavior, thought. We have no inkling of the internal factors that preside over the restitution of forms by regeneration, or of many other matters, including the immanent finality exhibited by every animate system which we call a living being.

Vitalism is dead, to be sure, but this does not mean we know all there is to be known about the phenomena of life. Living creatures are to be seen as consortia of organs whose functions mutually harmonize, complement, retard, stimulate one another. As physicochemical phenomena they do not escape universal determinism. Their parts are arranged and adjusted to create and maintain a new state, life. The living creature is not at all passive, and to every variation in its external environment or internal composition, it opposes a reaction that is not randomly chosen but tends to restore it to the conditions in which it can go on living as before.

Immanent or essential finality should be classified among the original properties of living beings. It is not to be disputed, merely acknowledged as existing.* Those who deny it—Rabaud, Matisse, and other biologists—have a great deal more to say about transcendental finality; their criticism strike home in the case of the excesses of which certain Darwinians are guilty, whose devout finalism is as like Bernardin de Saint-Pierre's as two peas in a pod. Put "Natural Selection" where he said "Providence" and you cannot tell the difference. The fact is that Darwinian logic taken to extremes proposes and imposes a totally optimistic view of nature worthy of Voltaire's Dr. Pangloss—every creature has its own useful niche, every organ or function is the product of a selection, through which it attains ultimate perfection.

Despite the equilibrium of fauna and flora and the high degree of adaptations, all is not quite so lovely. Failures and misfits abound; we have already discussed them and could go on doing so for a very long time.

The Darwinians have coined the terms pseudoteleology and teleonomy to designate the finality which they at the same time deny. Appearances are deceptive, they say; the materials of life are always the work of chance. What some take for finality is only the result of the ordering of random materials by natural selection. Even were this to be true, as it is not, the demon of finality would still not have been exorcized. For natural selection is, in essence and function, the supreme finalizing agent.

Actually, the terms pseudoteleology and teleonomy are the homage paid to finality, as hypocrisy pays homage to virtue. Giard (1905), himself a shrewd scholar but blinded by a foolish anticlericalism, went so far as to abjure Lamarckism and write, "To account for the wondrous adaptations such as those we observe between orchids and the insects that fertilize them, we have hardly any choice but the bare alternative hypotheses: the intervention of a sovereignly intelligent being, and selection." He cannot have seriously subjected his supposed dilemma to critical scrutiny or he would have seen that he was substituting for the dethroned divinity just such another, a sorting and finalizing, in sum transcendental, agent, natural selection. Paul Wintrebert, a convinced and even intractable atheist, did not fall into the same trap but realized perfectly that Giard's alternative involves, whatever opinion be held, recognizing the intervention of a purposive guiding agent.

Giard's concept, which is that held by many atheists and freethinkers, gives a singular and belittling idea of God. The Almighty,

---

*Man is a maker of finality and by his tools and machinery prolongs, if you will, the finality of his own organs, driving it beyond its natural limits.

obliged to remodel and retouch His own handiwork all the time, is baffled by obstacles His omniscience failed to detect. He is not even a demigod, but a mere pawn, a vague deity designed for crooked-thinking scientists.

Nature has its laws. The determinism of the things that flow from first causes suffices to explain the phenomena occurring in the material universe, whether it be made of inert matter or of living things. Let us not invoke God in realities in which He *no longer has to intervene*. The single absolute act of creation was enough for Him.

Evolution, and the world of living things, place obstacles that we cannot clear in the way of our thirst for understanding. Let us still go on trying, and we shall end by understanding. Giard's dilemma, adopted unreservedly by the Darwinians, is a false one, so we need not take any notice of it.

With more modesty than the Darwinians, we do not claim to know the author of that finality, which in fact goes far beyond the putative work of natural selection since we encounter it at every level in every living creature. Life is seen to be dependent on the finalization of a complex and architected physicochemical system, which still does not exclude next-best things, or even failures.

Without quibbling about its nature and justification, let us note that, far from abolishing or eluding determinism, immanent biological finality implies one that is strictly channeled and utterly opposed to chance.

The notion of physiological function implies the performance by an organ or system of organs of something precise, definite, constant, on which the living creature's survival depends. If the function is not performed, life stops. Immanent finalization is, in our opinion, self-evident.* Claude Bernard and most other physiologists have shared this certainty and were called finalists for their pains. So what? Since when has it been wrong to establish truth?

Immanent finality is an *intrinsic property* of all living creatures; without it they would not exist. Regarded as autonomous functional units, their component parts (organs, tissues, or isolated cells), just as much as any other property (feeding, defense of the organism, growth, and reproduction), are subordinated to an end. In the case of these properties, there is no argument, but just pronounce the word finality and every biologist is up in arms, most likely because biologists do not distinguish between *de facto*, immanent, and transcendental finality. On the latter they have

---

*Functions do not accommodate random variations; physiological disorder creates pathological effects and ends in death. Functional deviation is for the future of the individual and the line more important than formal variation.

little or nothing to say; it is a matter of metaphysics. Nor do they have to elaborate the meaning of living things in the macrocosm; this is outside their field and involves a real or putative finality which eludes the biologist's grasp.

Immanent finality manifests itself in life itself, since this always tends to preserve itself and propagate. If such a tendency is not a finalization of physicochemical phenomena, then there is no meaning in the word *end*.

Finality would be no more shocking to our susceptibilities than assimilation, growth, reproduction, or regeneration if we did not overload it with some sort of vague mysticism or vitalism with which it has nothing whatever to do. In the Darwinian system, the living creature is passive; it undergoes, but has no active part in change. Consequently, *immanent finality cannot come from within* but is introduced from the outside by a transcendental agent, natural selection, which acts on behalf of survival of the individual and, therefore, the species. Invoking necessity may, if you like, attenuate the Darwinian passivity of living things but, as we shall see later, it is not the main evolutionary device.

Some natural philosophers, including Matisse (1942), argue against finalism that there is no sign of finality in the world of the inanimate and that the living creature is itself no more than a sum of physicochemical phenomena. Thus they conclude there is no finality in the biocosm. Let us understand one another; the living creature is not a mere *sum* of physicochemical phenomena but, rather, much more nearly an integral of them. The architecture, the spatial and temporal organization, animate a whole which forms a coherent system of tightly coordinated parts.

Thus Matisse's error of reasoning is flagrant. You cannot compare two states—inert matter and the living creature—that differ profoundly but yet conform to the same laws of physics and chemistry. It is a worse mistake than comparing a stone on the road with the Château of Versailles. Life emerges from an organized system as electricity is produced in a thermal or thermonuclear power plant, except that life is not a particle of energy like light or electricity. It is action, achieved by having recourse to the greatest diversity of forms of energy.

The living creature, as an autonomous system with its parts rigorously integrated in a whole, born of another system identical with itself, functions by itself, with no need of any engineer or mechanic.

Death, the differentiator, and the bonus of survival: these are the tools of natural selection, which, according to the Darwinian dogma, is a reasoning entity responsible for maintaining the well-made, the properly functional, in the world of living things. It separates the useless from the useful, the wheat from the chaff. It determines the end of every living creature, of every population.

Here we are, up to our ears in transcendence: Any organization or function is controlled by it. The building blocks for making living creatures are utilized by it. Thus, the entire world of living beings is transcended by phenomena that create finality. Natural selection working for the continuance and welfare of animals, plants, and man himself, is seen to be the grand law which organizes the living universe.

So the Darwinians, who fancied they had exorcized finalism and transcendency but forgot to analyze critically the idea of natural selection, failed to see its implications or metaphysical consequences. They thought they were absolved from giving any finalization or deistic interpretation by decreeing that on earth all is but deceptive appearances; finality is a sham, guided evolution illusory. How is it possible to understand such an attitude? We cannot pretend that nature (with a capital or a small "n") copies man, the latest of its creations. So we are forced to admit, according to the Darwinian view, that nature acts blindly, unintelligently, but by an infinitely benevolent good fortune builds mechanisms so intricate that we have not even finished with analysis of their structure and have not the slightest insight of the physical principles and functioning of some of them.

The living creature, by its control systems, fulfills two requirements, adapting to circumstance and maintaining equilibrium and unity. But its capabilities of adaptation are not unlimited, as we have shown. The law of living may be expressed by an oxymoron: fluctuating in equilibrium. The paradox in only superficial. *Any living thing* has at the same time a system of normal functioning and balancing systems to offset deviations in the environment.

It follows that all creatures from the protozoan to the mammal have a control capability whose effects are of several kinds:

1. Immediately adequate reactions (autoadaptive, Lamarckian). Most regulatory phenomena, from compensatory hypertrophy to immunity, fall into this category. Every living thing is rich in devices and properties enabling it to counteract the blows it suffers, and the aggressions that it suffers, like the cells of the lung reacting to dust, the reactivity of the Malpighian layer of the skin to mechanical friction, of dermal pigmentogenesis to the effects of sunburn, and so on.

2. Behavioral reactions that are appropriate responses to sensory messages bringing news of the outside world. To each sensation or message the animal will normally oppose an automatic reaction warding

off the danger, procuring food, or bringing it into contact with its own kind.* Most behavioral control mechanisms are an integral part of the inheritance of the species.

These two categories of reaction are biologically finalized and *in no way aleatory;* regeneration restores the limb lost by the newt; the loss of a kidney causes the symmetrical remaining one to become hypertrophied; all the reactions governing internal balance, homeostasis, demonstrate this irrefutably.

The organism sometimes fails to respond adequately to some environmental variations or aggressive agents. The reaction may be inadaptive, or indefinite as Darwin used to say. The reactions of certain Crustacea (*Artemia,* etc.) to increased salinity, or of certain animals to the administering drugs, belong to this category. The regulating capability changes according to the species, or does not affect the same organs or functions. The ability of tailed amphibians to regenerate is highly developed, but in tailless amphibians it is feeble or nonexistent. The reactions of lowland plants transplanted to high altitudes have shown us how the control capability varies from one species to another.

Evolution's relationship to finality is multiple and sometimes hard to trace. The advance of evolution toward greater complexity, more psychic insight, and refinement is undoubtedly finalized. Where disorder and the unoriented reign, evolution ceases. As we have demonstrated, within each line a certain form, a certain plan of organization or *idiomorphon,* will tend to be achieved, determining failure or success. Biological finality is no all-wise Providence, watching over nature to prevent it from blundering and making mistakes. It is independent of what, by a value judgment, we call useful, harmful, or indifferent to the individual or the species. The animal or plant follows an evolutionary path and remains faithful to it, to its own salvation or damnation.

"Invention" in the living creature is never the product of a single mutation. It combines, relates, adjusts to one another, separate organs or parts of organs. It creates an arrangement that assumes a definite function, one integrated in the plant or animal unit. The new science of *bionics* is aimed at discovering such "inventions."

---

*Sensation is not always followed by an adaptive response; e.g., the case of tropistic reactions, the moth that is attracted to and destroyed by the flame, needs no dwelling upon. Tropism, or rather taxis in the pure state, i.e., not integrated into organized behavior (instinctive complex), is a phenomenon that seldom occurs in nature. When integrated, it becomes just another element adapted to a finalized conduct (positive phototaxis, toward the sun, in swarming ants and termites, which drives them out of their nests and covers lengthy distances); cf. Grassé and Noirot, (1951).

Every human invention gives tangible form to that which we call an idea or a plan, what would perhaps be more properly called an essence. *We are entirely ignorant of what takes place in biological invention.* No experiment can explain the notion of plan or project. The ontogeny and maintenance of the organism obey commands and mechanisms that are automatic and inborn, inscribed in the genetic code and inaccessible to us. However, behavior can be the subject of experiment, and in various cases its mechanism has been totally broken down into its component parts.

The fulfillment of a plant is not the result of a toss of the dice. We have to have recourse to an antichance, to the Darwinians' natural selection; we are convinced that there are other, more efficient ones working directly in the innermost depths of the cell. We have pointed out that in phylogenetic organogenesis, natural selection has to be prophetic if it is to act efficiently on the first crude attempts at the budding invention. If at every turn in evolution we invoke a happy chance, will not such a kindly deity end by turning a deaf ear to us?

The role assigned to natural selection in establishing adaptation, while speciously probable, is based on *not one single sure datum.* Paleontology (cf. the case of the transformation of the mandibular skeleton of the theriodont reptiles) does not support it; direct observation here and now of the genesis of a hereditary adaptation is nonexistent, except, as we have stated, in the case of bacteria and insects preadapted to resist viruses or drugs. The formation of the eye, the inner ear, of cestodes and the whale, etc., does not seem possible by way of preadaptation. Besides, paleontology teaches that the evolution of the stirrup bones of the inner ear took place exceedingly slowly by the unambiguous addition of tiny changes, preadaptation had nothing whatever to do with it.

The role of natural selection in the present world of living things is concerned with the balance of populations; it is primarily of demographic interest. To assert that population dynamics gives a picture of evolution in action is an unfounded opinion, or rather a postulate, that relies on not a single proved fact showing that transformations in the two kingdoms have been essentially linked to changes in the balance of genes in a population. Circumstances occasionally award a given mutation a selectivity bonus, but for a variable time, as witness the heterogeneity of populations due to the abundance of alleles of a single gene and their composition over time. Studies on natural populations in their own proper environment show that the composition of genes is changeable and that dominant species vary over time. Ford (1971), in his book, says precisely this and nothing else; as for seeking in it proof of the formation of new species, there is no such hope.

Population dynamics depends on the influence of the environment on the organisms, and the appropriate selection of genotypes. The gene composition of populations changes continually by the occurrence of mutants. For tens of millions of years, populations of *Drosophila* have undergone millions of mutations. What is left of them? The insignificant modifications discovered by laborious analysis by Dobzhanksky and Boesiger (1968). These tiny, disorderly fluctuations of genes start no new line; they are apparently unconnected with the great process that has given birth to types and subtypes of organization. In the chapters that follow the reader will learn what place and significance we attach to variations in the gene composition of natural populations. *No* experiment justifies the assimilation of demographic changes of population to a slice of evolution as an innovative, creative process.

VII

Evolution and Necessity

WHAT BIOLOGICAL NECESSITY IS: CONFUSION WITH USEFULNESS

The philosopher considering the Universe in its entirety is led to admit that there is only one necessary, absolute being, God. All other beings are contingent; this is why Pascal said of himself: "I feel that I might not have been . . . therefore I am not a necessary being" (*Pensées*, No. 597). The proposition applies equally to every living thing.

But instead of lifting up our eyes to the highest summits of metaphysics, let us keep them fixed on more accessible objects, on physical reality. In my opinion what is necessary is what cannot be otherwise—a concept much like Aristotle's and adopted by the philosophers Kant, Hegel, and John Stuart Mill: ". . . Whether for beings, happenings or truths, he who says necessity says the impossibility of the contrary."

Causal necessity, so much debated by thinkers, gets along very nicely with this definition. Every mathematical or geometric demonstration *per absurdo* consists in showing the necessity of one or more conditions. The geometric figure "triangle" is formed by three straight lines on the same plane, each intersecting the others. All these conditions are necessary. If one is not satisfied, there is no triangle. The existence of the mathematical object is tied to an absolute necessity.

Biological necessity takes a different form, and in recent times several concepts have been confused under this term. Life is an epiphenomenon arising from a complex, structural, and autonomous system forming an object endowed with an irreducible individuality. Certain conditions stemming from the chemical nature of the constituents, the arrangement in which they exist, and the functions they serve are necessary for the organized creature to live.

For the bacteria they are as follows: proteins, including enzymes, DNA, RNA, ATP, membrane; for the cell, the same plus a nucleus in which the macromolecules of DNA, in association or combination with sundry proteins, form chromosomes, mitochondria with DNA, Golgi apparatus, ergastoplasm. In plant cells there are added plastids impregnated with chlorophyll and containing DNA. This leads to the proposition that in any living being there coexist structures, chains of chemical reactions, some basic and some accessory. The distinction is somewhat arbitrary, no doubt, since both kinds are united in any plant, and any animal. But it has the advantage of indicating what is immutable or necessary in the living creature and what is transitory and contingent.

The notion of biological necessity is further classified if viewed from an evolutionary standpoint. It is certainly a truism to say that the first living thing, formed of prebiotic materials, fully satisfied the "necessary" conditions for life, conditions immutable and fulfilled in both kingdoms. There is no harm in repeating this self-evident truth, because it shows where genuine necessity lies, the one which is a *sine qua non* for life.

Once biogenesis occurred, evolution no longer dealt with the absolute but rather the contingent; in other words, *what was useful took the place of what was necessary*. Some biochemists who have written about evolution have failed to understand this. In all living things we distinguish three categories of characteristics: the necessary, the useful, and the indifferent (harmful, on occasion). The first is absolutely indispensable, the second contingent and inconstant, and the third scattered irregularly among the species populating the various environments of our planet.

Thus, against an immutable and necessary backdrop evolution has diversified the two kingdoms *ad infinitum*. As to whether the diversity of plans of organization and forms was useful or not, that is something else. It enabled the various environments to be conquered . . . but did it? Bacteria and Phanerogamia are found everywhere, regardless of the uniformity of their structure.

Let us go a step further. What need do reptiles have for a secondary palate, a mandible reduced to a dentary only, suitable for mammals? Lizards, snakes, and tortoises have gone on living with no partition in their buccal cavity, no complexly structured mandible. Besides, the palate is found, with no other premammalian characteristics, in the crocodile, which is certainly out of place in the genealogy of mammals.

Thus we see the difference, biologically speaking, between the necessary and the useful. *There was no necessity for theriodonts to acquire a secondary palate*, which really served a purpose only in the case of mammals by creating in the splanchnic skull two superimposed and separate stories, one for respiration and another for food.*

The orthodox Darwinians (Haldane, Ford, Julian Huxley) have avoided confusing necessity and utility (a pitfall as dangerous for the biologist as for the philosopher). They practically never mention the former, which enters into the question only where fundamental biological functions are concerned.

The Darwinian doctrine implies that selection is "motivated" by the need the living creature feels to acquire, or modify in itself, something by which its condition is *ameliorated*. Need as a possible cause of evolution enters the picture whenever the creature finds itself in a state of imbalance with respect to its environment, external and internal. A change in the environment causes in the plant or animal a discomfort, the removal of which requires the initiation of a control mechanism. If by misfortune the latter is missing or inadequate, the creature dies. The control mechanisms ensure that needs are satisfied before being felt. Such a need does not exist for the preadapted mutant which finds itself already attuned to the modified environment. In that case evolution occurs without the impetus of necessity, now conjured away by chance alone.

We may also relate need, not to the individual, but as the Darwinians do, to the population upset by a change in the environment. To survive,† such a population needs to contain in itself variants (mutants) preadapted to the new environmental conditions.

Selection weeds out from the population the "needy," or "necessitous" individuals, and keeps the fittest, the balance and stable. The role of necessity seems primarily a negative one. It is the consequence of random preadaptation that without any law or rule meets the needs of the population. Those that perpetuate the species, that evolve, are in this view the random preadapted individuals—they and they alone.

Under the Lamarckian view, as amended and stated by Wintrebert, need assumes quite a different character. It stimulates the organism and compels it to change. This is what he expresses by a striking image used as the title of one of his books, "The living creator of its own evolution." For the time being we shall not examine how right or wrong the notion

---

*In the case of the theriodont we might speak, as Cuénot did with respect to other animals, about *prophetic organs*, whose genesis is chiefly due to internal factors.

†Here, too, let us clearly state what we are talking about: More often than not it is a question of utility, not necessity, as the heterozygotic state of natural populations proves.

may be. We merely note that it assigns a positive role to need in evolution whereas Darwinism maintains that every biological novelty is born of chance; "necessity," or what is claimed to be necessity, causes those experiencing it without being able to cope with it to be rejected.

Some animals and plants can satisfy their most fundamental needs in various ways. We have already mentioned "mixotrophy" in green flagellates and certain algae. Thus the cell needs oxygen to breathe and fulfill its chemical cycles. And yet a few species, when placed in an environment devoid of oxygen, get out of their predicament by making use of a fermentation reaction enabling them to carry out the essential oxidations with no gaseous oxygen, and so find the energy they cannot do without (insects). Higher plants placed in an oxygen-free atmosphere, provided they have enough carbohydrates either in themselves or in their environment, make use of alcoholic fermentations (*Raphanus sativus,* studied by Molliard, 1907). Yeasts and numerous bacteria are aerobic or anaerobic according to whether the medium contains free oxygen or not.

The living creature does not respond to need in a uniform manner, but has recourse to various solutions. There is nothing like this in the mathematician's or the geometrician's "necessity," since mastery over their own axioms also makes them masters over the necessity governed by these axioms.

The philosophical structure, if you will, of the living thing accounts for the variety of its responses to changing circumstances. For, in its organic and functional complexity, it forms a mosaic in which the necessary shares the border with the contingent, the useful, the indifferent, the harmful. The genesis of such a constitution is not explainable by any unambiguous causality.

To this scale of "values" should be added "luxury" characteristics: colors or adornment on which neither the existence of the individual nor of the species depends. The diversification through evolution of living things is assuredly not determined by any necessity. It may be argued that it is the consequence of an exact adaptation of every species to a biotope, a specific niche, but such adaptation does not, by its always relative nature, correspond to the absolute point of the necessary.

The argument calls into question the reality of sympatric species; and yet these are commonplace in large homogeneous environments. Copepods, amphipods, etc., living out at sea have genera rich in species, very often caught in the same sweep of a plankton net. Similar situations are observed in prairie, pampas, and steppe populations. The genesis of sympatric species is not easily attributable to necessity alone.

The Darwinians are extremely hard pressed to find an explanation for the presence in one biotope of two, three, four, or even more species of the same genus. Theoretically, different genotypes do not suit a single given medium. Between sympatry and the axiom of survival of the fittest there is an antipathy.

NECESSITY AND PERSISTANCE OF "STOCK FORMS"

Evolution as revealed by fossil remains of plants and animals does not bear the characteristics attributed to it by theory. From one parental stock we get variants that are perpetuated in their offspring in one or more lines, *but in numerous offshoots, classes, or orders the original stock or types also persist.* This raises the following question: What necessity is there for the stock to vary since it flourishes and has persisted in its unvaried form from the most ancient times? Relict species insistently pose the same question. They cannot have been so badly adapted as is imagined, since they have endured.

Sometimes the ancestor cohabits with its own progeny: The rectal pouch of the lower termites carries a fauna of large and complex flagellates that digest the wood ingested by the insect and act as symbionts. In the same pouch there swim side by side *Hexamastix, Trichomonas,* several related genera, *Foaina, Devescovina,* etc., and large-sized species: *Trichonympha, Joenia, Pseudotrichonympha, Staurojoenina,* etc., of very elaborate structural complexity. Zoologists agree in considering *Polymastigina* to be the common ancestor of all these genera and still others not mentioned here. These flagellates, which share a common ancestor (*Eutrichomastix, Trichomas*) and inhabit the same environment, have differentiated into seven or eight lines that have evolved in *very* different and perfectly well-defined ways.

In other parasitic faunules of protozoans or nematodes, in which *all* individuals are subject to the same conditions, the archaic species are just as flourishing as the more complex but less ancient ones; they are even more resistant to lack of food and water. The differentiation of orders, genera, and species took place without selection having to intervene.

The paleontological evidence raises practically the same problem as the evolutionary evidence. For example, one and the same stock of insectivores gave rise to the Tarsioidea, Lemuroidea, simians, and a few other suborders that died out in the lower Oligocene. *These lines cannot have been subject to the same necessity,* or they would not have been different. Hence the dilemma: Either necessity does not come into it, or the stock is nonhomogeneous and divided into isolated populations, each responding to its own necessity. This is tantamount to asking the follow-

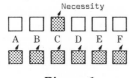

Diagram 1

ing question: Can the same necessity be responded to in different ways, or was there no necessity working on the evolving organisms?

Darwinism is uneasy in the answers it gives. It does not explain how several lines are born from the same stock. What has to be assumed is that the stock carried within itself as many necessities as there have been lines which have emerged to respond to them.

The Darwinians will not fail to retort that the same stock may from time to time have experienced separate necessities, thus the multiplicity of the lines successively issuing from a common ancestry. Of course, the biologist can give free rein to his imagination, but in that case of what value are the explanations given?

We have to recognize that the stock may have two different kinds of relationship to its offspring: either (1) it persists intact or practically unchanged and pursues its fate whereas the new line(s) it has engendered continues to evolve in its own patterns; the opossum (*Didelphis*), which, with a few insignificant differences of detail, is the double of *Eodelphis,* which lived in the upper Cretaceous (Alberta, Canada), displays a vitality surpassing that of the more evolved South American or Australian species; or (2) it alters and is insensibly continued along its new line of descent; this appears to have been the case for the theriodont reptiles which, tiny change by change, evolved into mammals.

Assuming that necessity was the driving force of evolution in either case, it follows (1) that if the stock persists, necessity existed only for one population of the ancestral species (top of Diagram 1); and (2) that if the stock vanishes, the necessity applied to all the individuals representing it (bottom of Diagram 1). But what is most important here is that in a few cases the "stock forms" coexist with their modified descendants. This fact contradicts the postulate of necessity.

NECESSITY, SELECTION, AND HETEROGENEOUS POPULATIONS

Various facts cast a doubt on the universality of the role assigned to necessity and selection in the evolutionary process.

Upon further consideration of the Darwinian postulate, how extraordinary is the fauna of a pond. In or on the same mud live bacteria, amebas, flagellates, nematodes, mollusks, insect larvae, etc. No doubt it will be argued that the biotope is not a simple one, but comprises several ecological niches, thus supporting a variety of inhabitants. This is hardly valid, for the pond's mud is a homogenous medium, its surface often covered by bacterial gels, is markedly different from the rest of it, but to a thickness of several centimetres the bottom layer of the pond water is homogeneous in its composition.

The cohabitation of species belonging to groups widely different in system teaches us that in one and the same environment, separate types of biological systems ensure the survival of one and all. Although built on totally different lines and not directly linked in kinship, the amebas, colonies of choanoflagellates, hopping *Bodo*, spirostomes, colonies of Vorticellidae and other ciliates, planarians, rotifers, nematodes, satisfy all their needs in the *same* environment. You would look in vain for the necessity which changed an ameba into a flagellate, a flagellate into a ciliate, a protozoan into a metazoan, and so forth.

All the animalculi of the pond vary and mutate like other species. The necessities compelling changes in the environment do not initiate the evolution, which ought, theoretically, to be convergent (responses to the same physicochemical agencies). Each group retains its own characteristics; what are claimed to be "existing necessities" prove powerless to induce their evolution.

From this we draw the conclusion that to maintain the living being, complication is by no means a matter of necessity. The animal or plant, be it uni- or multicellular, bears within itself the ability to persist. This property is already apparent in the cytoplasm of the bacterium, the simplest of all beings. Its control systems have enabled it to adapt to *all* terrestrial environments, including inorganic media, while still remaining itself a bacterium.

NECESSITY AND THE GENESIS OF THE MAJOR TYPES OF ORGANIZATION

When we consider the many kinds and species of blue-green algae (Cyanophyceae) and the multitude of bacteria and their successful establishment in all environments, we are may rightly ask what necessity, in the course of their long history, caused some of them to be transformed and acquire cellular structure.* Paleontologists think the schizophytes waited two billion years before evolving toward a differently patterned form; the birth of the cell is estimated at about 1.2 billion years ago. Since that time the schizophytes have not repeated their exploit, most likely accomplished by an insignificant fraction of them, assumed to have been affected by a putative but entirely unknown necessity. We have to imagine that a certain change in the environment, a precisely identifiable "cradle," created a necessity of an exceptional quality within a population of schizophytes involving the genesis of the eukaryote cell.† Everywhere else the "stock form" remained unchanged.

Bacteria and Cyanophyceae did not flourish less after the emergence of the cell, which did not bother them at all by its presence and even provided a new environment for them to populate. Although they have in the past undergone innumerable mutations, they oscillate within the akaryote design, altering without initiating any evolution whatsoever. By its unity of composition in all eukaryotes involving the same constituents (nucleus, centrosome, mitochondria, Golgi apparatus, ergastoplasm, hyaloplasm, etc.) and by its common fundamental cycles, the cell forces us to accept the idea of the monophyly of eukaryotes.

We have *no positive knowledge* of the evolution of the cell. It is presumed that soon after its appearance it differentiated into various types, each having its own mode of division (several in the case of protozoans and protophytes, one for the metaphytes, and only one for the metazoans) which made no change in its fundamental pattern. In the current status of biology everything supports the idea that the transformation of the schizophyte into a cell took place only once and that all existing uni- and multicellular creatures are products of it.

The historicity of evolution is affirmed by the fact of a unique "creation," but this hardly sheds any light on the fact that for such a vast period of time the cells born of the schizophytes underwent a tremendous evolutionary process whereas the "stock forms" simply went

---

*Actually, we do not know whether the cell originated with the bacteria or the Cyanophyceae; we have no paleontological evidence enabling us to decide between the two (see Chapter 1).

†It will be recalled that living beings are customarily divided into two main sets: (1) the akaryotes have no nucleus with a definite, chromosome-containing membrane; inside the actual cytoplasm they contain a self-enclosed macromolecule of DNA; the akaryotes have neither mitochondria nor Golgi apparatus; they comprise the bacteria and Cyanophyceae (often wrongly called blue-green algae); (2) eukaryotes have as their fundamental anatomical unit the cell, with a nucleus bounded by a perforated but clearly defined membrane containing a constant number of several chromosomes, with mitochondria enclosing one macromolecule of DNA, and with one Golgi apparatus.

through the paces of their tiny modifications while remaining authentic bacteria or Cyanophyceae.

Actually, in creative evolution, the existence of a driving "necessity" remains unproved. We do not know, and never shall know, whether a necessity impelled *Hyracotherium* to become a horse, *Moeritherium* to become a *Mastodon,* the amphibian a reptile, and the reptile a bird or mammal. Only a genius or a magician could, in the current state of science, given an irrefutable answer to these questions.

THE ANIMAL CREATES ITS OWN NECESSITY

Necessity is, we have said, the product of an anatomical functional disturbance creating an upset of equilibrium, physiological or anatomical, between the living being and its environment. Some neo-Darwinians imagine that at times the animal feels an urge toward a state differing from its own, which creates a fresh necessity: The living being creating its own needs.

Such a concept is an uneasy bedfellow for the doctrine. However, it is closer to the ideas of Lamarck and Wintrebert, who had the living creature play an active part in its own evolution. However, one Darwinian has also adopted the same interpretation. He writes: "If the tetrapod vertebrates emerged and gave rise to the marvellous flowering of amphibians, reptiles, birds and mammals, it was originally because a primitive fish "chose" to go exploring on dry land where, however, it could only move in clumsy hops" (Monod, 1971, pp. 142–143).

This entirely Lamarckian opinion is pure supposition, for there was no one to witness the fish's first walk on land or its transformation there into a walking tetrapod; we have no paleontological evidence to rely on.* It attributes to the animal a faculty of choice we are not prepared to admit.

As far back as 1809, Lamarck insisted on the important part played by changes of habit in the transformation of animals; he acknowledged that in such cases the initial cause of variation was not a change in environment but active intervention by the organism.

The changeover from aquatic to terrestrial life was probably preceded, in the case of rhipidistian Crossopterygii, by a long evolution preparatory to adaptation to the new environment and involving internal factors. It had to affect not an isolated characteristic, but the organism as

---

*So far no intermediate fossil between the rhipidistian Crossopterygii and the Ichthyostegalia, the oldest amphibians known, has been found. The absence of any evidence leaves the field wide open to conjecture.

a whole, since the variations had to be coordinated if they were to be meaningful and effective, and consequently could not depend on chance.

We are that much less inclined to accept the story of the little "Magellan of evolution" fish since the mudskippers *Periophthalmus* and *Boleophthalmus* very specifically reproduce its "experiment"; they hop about on the mud, climb onto the roots of the mangroves, and stand upright on their pectoral fins, as if on short legs. For millions of years they have lived like this and although they are hopping around all the time, however clumsily, their fins are still fins and do not turn into legs. How terribly unaccommodating of these animals!

Quite recently one of my colleagues drew my attention to the fact that *Periophthalmus* and *Boleophthalmus,* long accusomted to walking and skipping along on their fins beneath the mangroves, are completely satisfied to do so and will not at any price give up a tradition that suits them so nicely. Are they afraid that if they alter their ways they will get too big a boost from the pressure of selection?

The action of the environment, even when associated with a change of habits, cannot, to my mind, set evolution in motion unless certain internal factors come into play, as I shall describe in Chapters VIII and IX. Whoever regards necessity as the *raison d'être* of evolution is liable to slip into anthropomorphism. Another problem is that of detecting and measuring the needs of beings living millions of years ago in an environment of which we have only a few fossils and the sediments in which they are enclosed, to help us in their reconstruction.

What need did the diploblastic animals have to acquire a third layer? We can try to find out, since we can put down on paper whatever we like, but arriving at the truth is quite another matter. And these diploblastics, although engendering triploblastics, still continue with no sign of abatement. Even the usefulness of the change remains obscure.

Here we are back where we started from. From whatever angle one considers necessity as an efficient agent of evolution, its role is dubious, subject to the one reservation we have regarding the fundamental functions linked to the very manifestation of life (citric acid cycle, genesis of proteins, etc.)

NECESSITY-UTILITY: NOT THE PRIME MOTIVE OF BIOLOGICAL EVOLUTION

Our consideration of panchronic and artificial forms has shown the very serious difficulties which prevent us from assigning to necessity a determining role in evolution. There are many other obstacles to the theory.

Living things very soon acquired the control mechanisms they needed to experience environmental variations without injury. They were endowed with what was "necessary" for their own survival and the continuation of the species. What is acquired, aside from the fundamental molecular and anatomical structures and the primary functions, has no further link with necessity. All is contingent.

To meet what necessity did men acquire, in some cases, smooth hair, in others, crinkled, and, in yet others, woolly hair, or aquiline, snub, flat, or turned-up noses? Was there any need, in order for the perpetuation of the human species, for some men to have slant eyes, others black skins, for women to menstruate, for the infant to come into the world between the urinary meatus and the anus? To what necessity does the variable temperature of the sloths, the simple instead of double joint between the mandibles and the skull of mammals correspond? What necessity impelled the sunfish (*Mola, Ranzania, Masturus*) to reduce their spinal cord to the extreme (15 mm for a specimen measuring 2.50 m long and weighing 1500 kg) and not develop their caudal region? What advantage did they "expect" to get from it? Such questions are probably nonsensical, for neither necessity nor selection seems to have played any part in the genesis of these weird teleosts in which the exceedingly archaic heart and aortic arch, like those of the selachians and chondrosteans, is related to highly specialized characteristics.

Evolution went on; the necessity inherent in the achievement of the living creature was fully satisfied. But who shall tell us what necessity there was for life to appear on earth at all? This question is not addressed to the biologists, for it concerns the transcendental: let the philosopher or theologian answer it, if he can. If we invoke chance as a participant in the genesis and evolution of living things, we shall have to find some other trigger than necessity.

In a biological system, necessity appears the ineluctable consequence of a finality. Doubtless, it does not finalize vital phenomena any more than geometric or mathematical properties finalize a volume or an area; but integrated in them, it flows from the computer program contained and implemented in every living thing.

VIII

Activities of the Genes in Relation to Evolution

INTRODUCTION

Whatever has been said or written about the cytological mechanism of evolution is necessarily hypothetical, since the biologist has no clues to the past history of the cell and, as we have shown, cannot with any certainty isolate the evolutionary process in the present. Nevertheless, the cell is the crucible of evolution; in the final analysis, evolution results from a change affecting the nature, design, and mutual arrangement of proteins and nucleic acids, as well as from the acquisition of new segments of DNA (new genes) and protein molecules.

Novelties, and not merely modifications of preexisting genes, must take their place in the heritage of the species as represented, according to current biochemical thinking, by chromosomal, mitochondrial, and plastidial DNA's. Thus, interspecific affinities are given tangible form in this heterogeneously structured substance.

In every living thing a number of genes, those pertaining to the primordial functions, for example, come from the oldest ancestors (archeogenes). Those which determine the phylum, class, order, family, genus, and species are of more and more recent date as we go down the scale of magnitude of systematic units.

Since the genetic code which ensures the perpetuation of the structural, functional, and ethological characteristics of living things is written into the DNA molecules* and since the sequences of nucleotides

---

*The discovery of DNA, the genetic material of all living things from bacteria to mammals, is wrongly attributed to Avery *et al.* (1944). Actually, the role of chromosomes in

constitute wholes called genes corresponding to the achievement of some of the characteristics of the living creature, it ought to be possible, by quantitative and qualitative studies, to explore the gene complement of every living thing, and so determine its affinities with all others.

The genetic code is, as it were, the genealogical tree of the species, but so much has been merged, and there have been so many losses, admittedly offset by gains, that it is almost impossible to read it. The difficulty of its study notwithstanding, the science of heredity, or genetics, is remarkably advanced and precise. Among the facts it reveals, do any directly concern evolution and the birth of new characteristics? We cannot look for any without extrapolating from the present to the past, with all the hypothetical, indeed arbitrary, assumptions this entails. The historian explains the present in terms of the past; the biologist, in contrast, explains the past in terms of the present.

Prudence, the careful distinction of what is real from what is imaginary or interpretative, should be the biologist's constant concern when attempting to reconstruct the past, a very difficult undertaking.

EXPLORING THE GENOME

Genes and Mutations

Today we know that the sequence of bases in the double helix of the DNA molecule determines the primary structure of the proteins formed by transcription of the information carried by the messenger RNA (ribonucleic acid). Its constituent elements are sets of nucleotides each with one of four types of base (guanine, cytosine, adenine, thymine) pairs: adenine with thymine and guanine with cytosine. We know (cf. Nirenberg *et al.*, 1963; Speyer *et al.*, 1963) that every amino acid in a protein is encoded in a sequence, or *codon,* of three nucleotide bases. The possible combinations (by permutation of the four nucleotide bases in groups of three or codons) number 64, with only 20 kinds of amino acid.

The structure of the code is such that substitution of one base of a nucleotide by another, in first or second position in a codon, nearly

---

heredity was quite well known even before Avery started his research. The work of Sutton (1903), Boveri (1902, 1914), and many others had proved this. Since Feulgen's (1918) introduction into cytology of the detection of DNA by nuclear reaction, it was *perfectly* well known that DNA is the characteristic substance of the chromosomes of cellular creatures and the nucleoids of bacteria. Avery corroborated this but did not discover it. For the detailed history of work on chromosomes and the roles they play, see Wilson (1925).

always results in the use of a different amino acid, whereas substitution of a base in the third position produces a synonymous codon for the same amino acid.

We define the gene as the smallest segment of the DNA macromolecule capable of determining a constant characteristic. We know of the genes only as revealed to us by the mutation(s) they undergo. All invariants remain anonymous. This renders detection of the components of a genome exceedingly difficult.

Geneticists and molecular biologists are fully in agreement that the mutation or *genovariation* consists of the above-mentioned substitution of bases in the nucleotides forming the gene. Substitution and disordering of bases are phenomena that transform the gene into one or more others, known as its *alleles*. The mutation occurs wherever the DNA molecule replicates itself; the copy of the hemi-molecule is wrong at one point. To use a metaphor, every mutation is the equivalent of a misprint in a text.

According to the number of substitutions and their sites, breaks in the established order are followed by more or less marked effects. It is now known that some changes in the amino acids of a protein due to copying errors in the DNA have no effect upon the particular living creature. Work carried out on the molecule of myoglobin compared with that of hemoglobin has taught us that 80% of the homologous sites in the chains of polypeptides differ in their composition in amino acids, although both globins have very similar tertiary structures (folding of the molecule upon itself). Not all sites in the molecule are equally active; the most active are those that alter the least or not at all. Modifications of the other parts are of no consequence, but are *selectively neutral* mutations as described by Goodman (1961) and other biochemists.

Number of Genes

By the very principles of modern genetics it is agreed that a complex animal has more genes than a simple one. Concrete data on the subject are unreliable or even contradictory.

The genes of the fruitfly (*Drosophila melanogaster*)* are estimated at 10,000 pairs (some say 5000, others 15,000). Let U be the rate of muta-

---

*Beermann (1972) assimilates to a cistron (= roughly one gene) every light band–dark band set of the giant salivary chromosomes of *Drosophila* and chironomids. He gives an estimate of 5000–10,000, too imprecise to teach us anything about them. He also reports that every band of a giant chromosome comprises approximately 20,000 pairs of nucleotides. The number of DNA fibers increases from one larval stage to another (endomitotic multiplication of chromosomes); the figure given by Beermann probably counts chromosomes belonging to the last larval stage.

tions for the entire genome, u the rate per gene, N the total number of genes; then

$$U = Nu \qquad \text{or} \qquad N = U/u$$

The questionable item is u, for the geneticist has no reliable means of measuring it; figures vary by a factor of 3, which is much too great, so that the numbers quoted are only very approximate.

The frequency of mutations in man has been established as a function of certain hereditary diseases (epiloia, retinoblastoma, aniridia, achondroplasia, partial albinism with deafness, Pelger's anomaly, and some others). The data are too sparse for any precise estimate of the mutation rate. Some estimate the number of genes to be practically the same in the fly and in man, *viz.*, 10,000 pairs!

The equality of numbers is surprising, very surprising indeed. Although we know the structural complexity of insects, it may be forgotten that man's is far greater. How can the same number of bits of information produce two creatures so vastly unequal in complexity? Admittedly we can easily say, as geneticists do, that the 10,000 bits of "human" information differ *qualitatively* from those of *Drosophila*, that one bit is of a higher value than the other. But no one has the slightest hint of whether such a hypothesis is true. It is still 10,000 pairs each, very few for building a human being.

If the "one gene, one enzyme" aphorism of Beadle and Tatum (1941) is correct, the number of genetically determined characteristics being higher for a mammal than for a fly, we have to infer that, in the case of the former, one enzyme produces not one but several characteristics. Some mutations caused by a single gene consequently affect, by a sort of ricochet, several characteristics (frizzle fowl, rhinoceros mouse, etc.), but we cannot be sure that a gene producing only a single enzyme directly determines several characteristics.

Enzymes, structures, arrangements of organs, and synchronization of organs and functions demand a gigantic mass of information. Each constituent part of the smallest organ depends upon several genes. Identifying all the genes or groups of genes (cistrons, operons) involved at all levels of organization in ontogenesis, the functioning and maintenance of parts, is a tremendously important matter, but one whose solution is beyond our grasp at present.

This review of our meager store of knowledge about the quantities of genes in the different species confirms that we do not know the absolute numbers of these determining elements, nor whether there is any direct ratio of number of characteristics to number of genes. All this baffles the theorists who would explain evolution in terms of random variations of genes revealed only, or almost only, by accident and freaks.

Quantity of DNA and Systematic Rank

Quantitative research on DNA has produced facts of great interest but not at all easy to interpret with the current state of our knowledge. The weight constancy of DNA corresponds to the numerical combination of chromosomes for a given species. But it must be remembered that in a very great number of plants and animals the number of chromosomes per cell varies, sometimes considerably, from one organ to another. The familiar process of endomitosis multiplies the set of chromosomes in a single nucleus retaining its membrane. Only the sex cells are not subject to variation in the numbers of chromosomes.*

We do not know what the consequences of the multiplications of these organelles may be. Bacteria, sporozoa, various plants (mosses, ferns) are haploid over long portions of their cycle, i.e., contain only one macromolecule of DNA (as in the case of bacteria) or a single set of chromosomes (n).† All other plants or animals are diploid; their chromosomes come only in pairs. Duplication encourages variation and makes sexual reproduction easy; moreover, any animal not having in its cells the $2n$ chromosomes for its species is, with exceedingly rare exceptions, not viable.

It would seem that the number of chromosomes, $2n$, or of their arms, has no evolutionary significance. We need only consult a list of chromosome sets to realize that often species in the same genus have widely differing numbers of chromosomes, the highest not always being a multiple of the lowest.‡ In other words, polyploidy (multiplication of all the chromosomes) is not the only mechanism causing the number of chromosome sets to vary.

The essential thing is the DNA content of the chromosomes, the genes. Nevertheless, some phenomena, such as the position effect, show that the topography of the genes along the chromosomes is not

---

*In some species "chromosomal" subspecies form. For example, the mole cricket, *Gryllotalpa gryllotalpa:* northern Europe 12 chromosomes, Romania 14, southern Italy 15. Closely related species such as *G. africana* and *G. borealis* have 23.

†We must bear in mind that any chromosome, of plant or animal, contains one DNA macromolecule associated with proteins including histones [nuclei of peridinians (protophytes) have none]. The ratio of DNA to mass of the chromosome is 15%. The ultrastructure of the chromosomes is unknown. The fixatives at our disposal, though numerous, do not distinguish it. The majority of work published on the subject describes artifacts and not realities. Our ignorance of the relationship of molecules of DNA, RNA, and proteins in the chromosome renders an understanding of the mechanism by which genes are activated difficult.

‡As an illustration, here are a few figures for the order Primates: *Lemur,* 44 to 60; *Hapalemur,* 54 to 58; *Galago,* 38 to 62; *Cebus,* 54; *Ateles,* 34; *Macaca,* 42; *Papio,* 42; *Cercocebus,* 42; *Cercopithecus,* 54 to 72; *Hylobates,* 44; *Symphalangus,* 50; *Pongo,* 48; *Pan,* 48; *Gorilla,* 48; *Homo,* 46.

without consequence for the properties of the genome. Chromosomal breakage followed by the adherence of the fragments onto other chromosomes in normal or inverted position, may or may not have consequences for the genic properties.

In-depth researches into Lepidoptera (Lorković, 1949), reptiles (Matthey, 1970), and rodents (Matthey, 1969) have revealed no generalized facts and teach us very little about the mechanics of evolution. Knowledge of the genes and their sites on the chromosomes and of their linkages and recombinations (chromosomal maps of genes) has enabled us only to define the differences among *Drosophila* mutants and among a few *Drosophila* species and gain some insight into the modes of transmission of apparently aberrant characteristics. Nor has it yielded more facts concerning the colibacillus (*Escherichia coli*), the fungus *Neurospora crassa*, maize, or *Oenothera*. In every case we cannot get beyond the field of strict genetics. So far, the locations of genes on human chromosomes have taught us nothing fundamental about our origins.

However, recent hypotheses concerning the phylogeny of primates have been based on the presence of bands more or less demonstrable by histological staining in the chromosomes of species presumed to be related (*Homo sapiens* and the anthropoids) (Lejeune, 1970; Turleau and de Grouchy, 1972; de Grouchy *et al.*, 1972). As techniques are perfected (we do not know whether the dye settles on the DNA or on the proteins of the chromosomes) and more numerous species and more numerous chromosome rearrangements are observed, we may get some worthwhile information. But studies and extrapolations concerning the shape and stainability of the chromosomes cannot supersede paleontological evidence, the arbiter of evolutionism.

Matthey (1969), an eminent authority on chromosomes, admirably sums the matter up: "... It is highly curious to note that major changes in the architecture of the chromosomes ... remain compatible with perfect stability of the genotype. These great mutations have practically no effect." But he adds judiciously that they do contribute to isolation of the sexes.

The recently announced revelation of two cryptic species of mammals (Gerboise monkeys) concealed by the name *Tatarillus gracilis* is edifying. In a single population on the same site in Senegal live individuals having 37 (male) or 36 (female) chromsomes, and others having 23 (male) or 22 (female). There is outwardly no means of distinguishing the two types. Hybridization between them seems to produce sterile individuals; hence the two sorts are probably sexually isolated. A fact such as this confirms that neither the number nor the arrangement of the chromosomes affects the characteristics determined by the genes, and only the presence of the latter has any importance (except in the handful of cases of position effect reported by geneticists). Matthey and Jotterand (1972), who found out the chromosomal disparity of *Tatarillus gracilis*, write that the morphological identity of the two kinds of Jerboa speaks "in favor of a relatively recent isolation of the sexes, since it has not yet been followed by chromosome mutations." Logic is on the side of this conclusion, but nothing can assure us that it is correct; there is no proof that it is true.

The fact that an additional chromosome is present (polysomy), or missing, generally has very harmful effects on the organism. In man, the anomaly called mongolism is due to the presence of an extra chromosome 21 (trisomy 21). Klinefelter's sexual anomaly is due to an extra X chromosome (44A + XXY), Turner's, however, being attributable to the loss of either an X or a Y chromosome (44A + X). Conversely, the chromosomic aberration 44A + XYY is accompanied by disturbances usually so mild that they go unnoticed. Nevertheless, it has been maintained that the extra Y causes increased aggressiveness or chromosome criminality; the statistics compiled so far should be treated with the utmost caution.

Thus, changes in the numbers of chromosomes other than by overall multiplications of the genome, or polyploidy, nearly always produce pathological effects on the organism concerned and do not appear to be among the primary factors in evolution.

As we have already said, the quantity of DNA in the genome should be proportional to the number of genes. More complex organisms, that is, those requiring for their ontogenesis and functioning the largest number of genes, should be the richest in DNA. Actually this is hardly apparent. The nuclei of many protozoans and protophytes and of the lower metazoans contain just as much if not more DNA than those of birds and mammals. Tunicates contain 0.3 pg* of DNA per nucleus; agnathans (= Cyclostomata) 5.00; chondrostean fishes 5, 5.6, 7; Actinopterygii 0.9 to 6.1; Dipnoi (*Protopterus, Lepidosiren*) 100 pg.

The urodelan amphibians also have a high DNA content: *Ambystoma tigrinum* (axolotl) and *A. laterale* have 102 to 105 pg per nucleus (MacGregor and Uzzel, 1961). *Amphiuma*, which forms a whole family of its own, would seem to have broken all records for DNA by weight, with 168 pg per nucleus; in two of the plethodontids, in contrast, the DNA content is only 20 pg per nucleus.

Anuran amphibians are much poorer in DNA; the frog (*Rana*) has no more than 2.8 pg per nucleus. Present-day reptiles have a DNA content varying from 2.8 to 5.3 pg: alligators (*Alligator*) have 5.0 pg/nucleus.

---

*The picogram (pg) = $1/10^9$ mg.

Birds have low DNA contents, varying according to species from 1.7 to 2.9 pg. Metatherian mammals (marsupials) have 8.8 to 9.9 pg (*Perameles*) whereas eutherian mammals have 5 to 7.1.

From these crude figures no conclusion can be drawn concerning the zoological origin or affinities between classes. But two questions arise:

1. Why isn't anatomical and physiological complexity always accompanied by increased DNA content?
2. What are the causes for evolved forms having less DNA than archaic forms?

In a few cases polyploidy or polysomy is responsible for the increase in nuclear DNA. This is practically certain in the case of many varieties of cultivated plants (apple, tobacco, iris, etc.) and a few animals. In amphibians with a very high DNA content, the study of karyotypes reveals no polyploidy.

It has been shown that the quantity of DNA is increased through repetition of certain genes, or *gene redundancy*, which was discovered during analysis of the mouse genome by the DNA hybridization technique (Waring and Britten, 1966). Schildkraut *et al.* (1961) had demonstrated that separate strands of purified DNA "recognize" one another and apparently reform associations according to their gene composition; the best way of producing the phenomenon is in an agar gel (Bolton and McCarthy, 1962). Hybridization has been obtained between DNA's of various bacteria and DNA's from different mammals, the association being just as fast in both cases, although the number of bases in the composition of the polynucleotides is said to be much higher for mammals than bacteria.

It has been observed that in the genome of various species there is a relation between the repetition of certain nucleotide sequences and the speed of association of the DNA's. The $C_0 t$ formula proposed by Britten and Kohne (1968), in which C_0 is the initial concentration in moles per liter of DNA nucleotides and t the time in seconds for forming an association, enables an estimate of the redundancy in a genome or one of its fractions. Redundancy is very common in eukaryotes from protozoans to man; the parts of the genome affected represent in ambystomid Caudata 20% to 80% of chromosomal DNA, which in the axolotl is reduced to slightly over 20 pg of DNA of noniterative genes. This is still a very high figure. Among mammals the average rate of redundancy is of the order of 30% to 40% DNA.

Two types of redundant sequence are observed. One represents about 10% of the total DNA and consists of sequences comprising roughly 300 nucleotides, repeated about 1 million times; heterochromatin, forming the satellites of certain chromosomes [e.g., mouse, guinea pig, shrew (*Microtus agrestis*), man], has just such a high redundancy rate; the remainder, about 20% of total DNA, consists of "families" of different sequences repeated 100 to 100,000 times.

The homologous segments are the ones that reassociate. What role does gene redundancy play? We do not know, but two American scientists, Britten and Davidson (1969), attempted an explanation by a complex model that interferes during cell differentiation as a regulator of the output of useful DNA's according to the stage of development and cell sites. The theory has no direct bearing on our topic, which is the mechanism of evolution and not normal gene activity in multicellular organisms.* Incidentally, it cannot apply to unicellular organisms in which there appear to exist several orders of redundancy.

Nonrepetitive genes (only one per chromosome) are apparently the most numerous, and genic regulation does not seem to be any less in the frog (*Rana*), whose genome has very little redundant DNA, than in the axolotl, rich in DNA. So then, is not the regulatory role attributed to redundant DNA pure fiction?

A gene repeated a million times ought to have a million more chances of mutating than a single one. Observations to date do not support this "logical" opinion.† Remember that the heterochromatic segments of chromosomes containing the redundant genes were, until quite lately, considered genetically null, or almost so.

Some cell biologists have formed the hypothesis that in redundant DNA the nonidentical families of nucleotides are the oldest, having accumulated as time has gone on mutations affecting the order of bases. The starting point of their differentiation would be these mutations.

Having reviewed the known facts, which are few in number and not always reliably established, it seems rash to assign a precise role to redundant DNA. The qualitative factor in all this may outweigh the quantitative.

When Watson tells us that 85% of the DNA contained in the chromosome set of a species is inactive, he is making a very rash supposition. No one can boast he knows all the properties and functions of this

---

*Britten and Davidson's proposed model is purely hypothetical in structure and role. It is possible to propose another, equally logical and likely. Such "flights of fancy" would do no harm if some biologists did not take them for positive truth. The chapter by M. Goodmann *et al.*, Evolving primate genes and proteins, in the collective volume "Comparative Genetics in Monkeys, Apes and Man" is written evidence of this confusion between science fiction and reality.

†Mutants are interpolated in the redundant sequences, so that the original genes and mutated ones coexist in the same series.

nucleic acid. He is well aware that the presence of $2n$ chromosomes in all cells is a vital necessity for practically all animals and the higher plants. The endomitosis so widespread in both kingdoms sometimes greatly increases the number of chromosomes. I have observed that in the excretory cells of the Malpighian tubules of Acrididae and crickets the number of chromosomes rises from the apex of the tube down. The degree of polyploidization seems to vary according to the functional status of the cell.

It may be that a certain mass of chromosomal DNA does not make up genes or cistrons, but no biologist in the present status of molecular biology can assert that it has no function in the life of the cell (see Commoner, 1968).

The Notion of the Operon

Among the latest concepts of molecular biology there is one that provides matter for thought. Instead of the isolated gene being the agent of the genesis of a character and a subunit of the individual, we postulate a set of genes (sometimes termed cistron) with acts together and which forms a functional unit.

The most complex such set is probably the *operon* (Jacob and Monod, 1961; Monod and Jacob, 1961). It corresponds to a segment of the DNA macromolecule and includes several genes coordinated in regard to their actions and effects.

This complex was envisioned following studies on the production of enzymes by bacteria of the genera *Escherichia* and *Salmonella*. *E. coli* metabolizes the β-galactosides by means of three enzymes: β-galactoside permease, β-galactosidase, and β-galactoside transacetylase. The first two are synthesized in equimolecular quantities, the third in a less simple relation. In other words, syntheses of these enzymes are coordinated. An operon is seen as several adjacent and functional genes (structural genes) plus one or more operator genes and a repressor. The operator acts upon the repressor (directly or indirectly) and thus regulates the quality of the enzymes produced (three enzymes in the case of the β-galactosidase). The β-galactosidase catalyzes hydrolysis of the lactose to form glucose and galactose.

What interests us in the operon is the necessary link between the structural genes and those which control them. It is unlikely that every structural gene is integrated with an operon, but it may be so in a great number of cases.

The operon linking a producer and a control system expresses an order. It cannot be created by a mutation, because, so far as we know, it does not have the ability to construct. It introduces a disorder into the arrangement of bases (adenine, guanine, thymine, cytosine), the only part of the DNA molecule for which disorder is not *ipso facto* destructive, and that part on which chance exerts its influence. For the genesis of the operon we must visualize an organizing antichance. No doubt for the Darwinians this can only be natural selection. And yet in such cases it is not necessary merely to sort out, but above all to organize and articulate with one another, spatially neighboring genes, and to coordinate their actions. Did these genes play any part in the life of the cell before they formed a coherent and operational unit? Whichever way we turn, we are confronted and swept off our feet by hypothesis.

No one seems to have any clear idea of what a mutation in the operon can be; does it directly alter the structural genes, or touch only the other components? Theoretically there is no reason why both should not be true.

It cannot be affirmed that the mutation of a gene in an operon is equivalent to that of an isolated gene, for repressive or other action on the operon is applied to the mutated gene, which also depends on the adjacent structural genes. It is not entirely imaginary, for it has been found that mutations, especially deficiencies, of genes contiguous to the operator render the whole operon inoperative. Restitution of the affected part restores the operon to its premutation activity (Jacob and Monod, 1961).

Variation of an independent gene is directly and immediately governed by natural selection, whereas the mutation of a gene belonging to an operon has, first of all, to be "accepted" and "tolerated" by the operon. This is an important point that needs to be answered by the Darwinians.

The Age of Genes and the Past History of the Living Organism

Every vertebrate, for example, gets some of its genes from its remotest ancestors. Most likely the fundamental biochemical phenomena common to all living things are due to the same genes (archeogenes) born of the same ancestors who in this way go on living in their descendants. This view has been supported by observations made by biochemists on respiratory enzymes (cytochromes), the enzymes and other materials involved in the citric acid cycle (or Krebs cycle), the hemoglobins, and various other proteins. The constancy of the functions of archeogenes, in bacteria as in man, testifies that they have changed but little throughout the ages.

Studies on embryogenesis have shown that such genes do interfere, and have in some measure enabled their phylogenetic affinities to be evaluated. For example, it has just been shown, using the organotypic culture technique, that the embryonic mesenchyme (liver and lung) of the bird's embryo is able to induce differentiation of the liver when grafted onto a rodent embryo. The genes involved are probably identical, or nearly alike; they must have existed in the ancestral reptiles that in different lines produced the stock of mammals and birds.

In a few lines the archeogenes have, however, varied or been replaced; thus, the mitochondrial genes of anaerobic organisms are not those of the aerobic ones. The mitochondria do not have the same structure or bear the same enzymes as those of aerobic species. In those of flagellates symbiotic with termites, strict anaerobes, no partition or tubule can be seen; the outer membrane is simple, an easily demonstrated macromolecule of DNA.

Any animal or plant has the genes of its phylum, class, order, genus, and species. Through them it receives tangible form. During embryogenesis the first characteristics to appear are the most general, those of the phylum, and the last are the most particular, those of the species. This order repeats in rapid succession that of the acquisition of the genes during the evolution of the line.

Research, not on the archeogenes themselves but on the proteins encoded and determined by them, has not yielded the phylogenetic information expected. Such has been the case of the "paleoproteins," the cytochromes. The cytochrome c of man differs by 14 amino acids from that of the horse, and by only 8 from that of the kangaroo (*Macropus*). Similar facts are found in the case of hemoglobin; the β chain of this protein in man differs from that of the lemurs by 20 amino acids, by only 14 from that of the pig, and by only 1 from that of the gorilla. The situation is practically the same for other proteins.

Some chemical analyses of primate genotypes (DNA) have recently been made (Kohne, 1970). The results are not uninteresting but are difficult to interpret. The numbers of nucleotides are so different from one genus to another that no intergeneric affinities can be perceived.

Quiescent Ancestral Genes

Atavistic Polydactyly in the Equidae

Unusual, not to say abnormal, facts cause us to surmise that certain genes that were phenotypically present in the ancestors persist in the genotype of the species existing now but no longer induce the formation of organs or parts of organs, and command no function.* Such are the various forms of polydactyly in the Equidae. They recur with varying degrees of expressivity. Sometimes the only trace of the extra digit is the nail or a horny fragment. More rarely, digits II and IV are developed with all their joints, and end in a small hoof, but since they are longer than digit III they do not reach the ground; in that case the three-toed foot imitates that of *Hipparion.* It is assumed that the genes determining the extra toes must have been latent in the ancestor† and have resumed activity by an unknown mechanism.

The genes determining the metacarpals and metatarsals II and IV have not vanished, since these bones remain in the form of nonfunctional stylets. The top joints, on the contrary, have left no trace, even in the embryo, which I have vainly examined for their cartilagenous or mucous anlagen.

The mutation by which digits II and IV reappear might be due (it should not be forgotten that in this matter any interpretation whatsoever is pure hypothesis) to a change in a single gene that removed the inhibition on the genes (which must be numerous) determining development of digits II and IV, but it is much too convenient to have at one's beck and call that obliging little imp, a gene.

Geneticists commonly invoke such helpers. In bacteria and *Drosophila* they have observed mutations which maintain a "premutation" type and have called them *reverse,* or in French "inverse." They believe that these mutations modify and inhibit the suppressor genes located in the same cistron and thus enable the genes determining the characteristics to operate. Maybe also, for a reason unknown to us, the ever-present genes governing organogenesis of digits II and IV undergo a sudden surge of expressivity. The single-toed state of the genus *Equus* is probably, and, at any rate, partly due not to the disappearance of the genes controlling digits II and IV but to their practically total inhibition.

When facts do not reveal their law, hypothesis reigns supreme; but a sadly feeble ruler it proves to be.

---

*Kohne (1970) proposes that some sequences of DNA fulfill no function. For reasons not apparent to us he thinks that since the sequences only underwent neutral (?) mutations, they did not cause selection. The same has been said of heterochromatic chromosomal segments.

†The reappearance of digits II and IV, or *atavistic polydactyly,* would appear to be unconnected with the doubling or tripling of toe III, likely to be due to a division of the embryonic anlagen of the toe (schizodactyly).

Tetrapterous Drosophila

Another teratological case likewise suggests the persistence of ancestral genes in the quiescent state. Four-winged *Drosophila* have sometimes been found in experimental breeding stations. The halteres, which are profoundly modified posterior wings, revert to wings, although imperfect ones. But the haltere is not simply an atrophied wing reduced to one or two veins; it is a complex sense organ that plays an important role in the balancing and control of muscular efforts. It contains numerous and various sensory receptors. Its function apparently requires, by reason of its structure, the intervention of several genes; it is probably the result of a long evolution.

The chief problem posed by tetrapterous *Drosophila* is to know what genes control formation of the posterior wings. *The genes of the normal wing are not those of the haltere*, whose sense organs are different from those of the wing in structure and function. It may be imagined that the genes controlling function of the haltere are in some cases the determinants of the normal wing modified in respect to expressivity, and in others, *new genes* controlling genesis of the sense organs appropriate to the haltere.

It may be, in this case too, that the tetrapterous state is restored by mutation of a single gene reactivating the wing genes and inhibiting those of the haltere. This would be a logical interpretation of the facts, but it has no solid basis. (The reader is referred to Chapter IX, dealing with the gene and its mode of action).

The difficulty is considerably increased when we realize that four-winged *Drosophila* has been produced in several breeding stations. To my knowledge there has been no study of them. The genotype of these monstrosities is likely to contain the ancestral genes (this cannot be the case for anterior wings changed to halteres) in a quiescent state. Parts of the same macromolecule of DNA, no doubt owing to their heterogeneity, exert an influence on one another through the enzymes whose genesis they control. Unless perhaps the cytoplasm constituting the environment does not affect expressivity, the penetrance of the genes is enough to inhibit any action by them.

The preceding facts teach us that hybridization of the intra- or extra-specific DNA's is not enough for exploration of the genic content of the species. Freak organs sometimes occur revealing genes that have remained hidden and inactive for millions of centuries in the heritage of the species and of the lineage.

Does this nullify, then, the paleontologist Dollo's law, that regressive evolution is irreversible? Yes, if the ancestral genes do persist and are kept quiescent by a gene-suppressor system, and if removal of the inhibiting mechanism restores them to their previous state. But no paleontologist has ever observed the reappearance of organs that have disappeared in the course of evolution; we are therefore entitled to think that reaction of the ancestral genes is an exceptional teratogenic phenomenon, having no effect on the evolution of organized creatures.

Heteromorphic Regenerations

As far back as 1897 Giard (1897) conceived the idea that regeneration may reveal some of the workings of morphogenesis. His intuition was correct. This phenomenon, which shows abnormal ontogeny or organogenesis, teaches us, upon close analysis, about the genes involved, notably in restitution of form.

We shall consider here the very particular case of heteromorphosis, the faculty of regenerating a different part of the body from the one that is missing, the regenerate reproducing some other organ. It was discovered by Loeb (1891), who saw a fragment of *Tubularia mesembryanthemum* (a cnidarian hydroid) regenerate a head at either end. Many similar cases have been observed: the formation of a head instead of a tail in an earthworm (Hazen), the formation of an antenna instead of a compound eye after amputation of an ocular peduncle in the crayfish (Herbst, 1896)—this decapod regenerates either an ocular peduncle or an antenna, depending on whether the section does or does not leave the optic ganglion in place.

Another *locus classicus* is that of the stick insect (*Carausius morosus*) studied by Schmidt Jensen (1913), Cuénot (1921), and Borchardt (1927): An amputated antenna is replaced by a leg. For heteromorphosis to be produced, the section of amputation has to pass through one of the two basal joints (I and II) of the antenna, or between the two (Borchardt, 1927). If Johnston's organ (normally housed in the second antennial joint) is partly or wholly retained, the regenerate will be an antenna, otherwise a leg. The heteromorphic limb is unarguably a leg, with femur, tibia and triarticulate tarsus (Cuénot, 1921). It is smaller than a normal leg and nonfunctional.

The insect's head is in six segments, three preoral, three postoral tightly amalgamated, and a small anterior region of nonsegmental origin derived from the embryonic acron, all forming a perfectly well-defined anatomical unit. Every segment except the first and third has its own pair of appendages, each having its own original use: antennae (segment II), mandibles (IV), maxillae (V), and labium (VI), and comprising two appendices joined together at their inner edge. The thoracic segments bear the legs; the abdominal segments have either lost their appendices or retain vestiges of them.

Antennae and legs already existed in the oldest insects (*Rhyniella* of the mid-Devonian). Since that far-off time the legs and antennae have been determined by different genes because these appendages do not have the same sensory receptors, nor the same musculature, nor the same nerve distribution.

In such a case regeneration does not arouse any sleeping ancestral gene but stimulates at an unusual site (that of the lost organ) the genes of another organ, which in normal ontogeny is formed at another point on the embryo. In heteromorphosis the organism makes a mistake; it does not call in the right category of worker for accurately repairing the lost organ, but rather others which build what they have been "trained" for, and not the missing organ.

Therefore, since we have no certitude that the antenna was preceded by an ambulatory appendage* and the compound eye by an antenna, we may wonder whether we are entitled to talk about ancestral genes unlocked by the removal of an inhibiting mechanism. The situation is quite different from that of the atavistic digit of the Equidae. What genes induce the formation in the stick insect of a foot for an antenna, and in the crayfish of an antenna instead of a compound eye? We do not know what the answer is.

It may be that the sensory cells, or those supporting the Johnston's organ, act on the cells in their immediate vicinity like an inducer–organizer. This is tantamount to saying that the cells serve as materials for appendicular regenerates when lacking a special induction of an ambulatory foot. The genes responsible for this organogenesis are the only ones to be stimulated by amputation of the appendage when Johnston's organ is missing.

"Creative" Regeneration of Nemerteans and Its Variability

Dawydoff (1942), in a dissertation that has remained practically unknown to biologists, made a detailed study of the reconstitution of the cephalic region in a nemertean, *Lineus lacteus*, that lives in the Black Sea off the Crimean coast. *A prebuccal fragment* of this worm regenerates a complete animal if it contains the whole brain and accompanying cerebral organs; to achieve this, the razor must cut a section immediately behind these organs. After cicatrization by closing the lips of the wound,

---

*Some anatomists rely on their intuition to tell them that the ancestral arthropods had one pair of undifferentiated but ambulatory type of appendices per segment.

epidermalization, and formation of a fresh ciliate tegument by migration and spread of cells from the lips of the cut, all the cells in the injured area and fairly remote areas are dedifferentiated, i.e., lose their own attributes and take on the appearance of embryonic cells with cytoplasmic extensions, in a confused mass. Each organ regenerates from what is left of it in the prebuccal fragment. If there is a bit of the esophagus, the dedifferentiated cells from it form a vesicle, by means of which the various parts of the new digestive tube will form. If none of the esophagus is left, regeneration of the digestive tube takes place at the expense of the mesenchyme. After 6–7 days a vesicle is observed whose cell wall is, to begin with, single-layered.

What is extraordinary about this regeneration is the variety of cellular processes by which it is achieved. Dawydoff counted eight for reconstitution of the digestive tube; the vesicle which is its anlage is derived (1) from the walls of the rhynchocoele (sheath of the proboscis), or (2) from the walls of the lateral vessels, or (3) from the mesenchymatous cells (intervisceral mesenchyme) without prior formation of a gastric cavity, or (4) by the same process but around a gap destined to become the aperture of the digestive organ, or (5) from a syncytial network (the term *syncytium* is used for a cytoplasmic mass containing several nuclei not separated by intercellular membranes) without prior assembly of the cellular material, or (6) from a syncytial network but through the intermediary of special cells (*gastroblasts*) differentiated inside the syncytium, or (7) from plasmodia, or (8) from undifferentiated and totipotent cells, the *neoblasts*.

Whatever process is used by the regenerate, *the result is the same*, a digestive tube and its various parts.

Without going into the details of the regeneration, plainly the cells building the new digestive tube belong with absolute certainty to the mesoderm. The eight kinds of regenerative modes (each having variants) make use of separate genes, i.e., theoretically, enzymes, chains of chemical reactions, differing from one kind to another. By causing the same individual to undergo successive regenerations, it has been found that later regenerations assume different modes.

Dawydoff was a great embryologist and trained observer never caught off guard, but having once described the variations of regeneration he returned to the interpretations and concepts of Driesch (1909), Gurwitsch (1914–1915), and Lubischev (1925) and attributed adaptability to circumstances and the variety of building processes to nonmaterial factors. I do not go along with this vitalistic view.

It should be noted that only Crimean *Lineus lacteus* specimens display great variety in regeneration of the digestive tube. Those from elsewhere do not have it. It is most likely a property linked to the presence of one or more particular genes.

Heterogenetic regeneration uses genes that remain inoperative in normal ontogeny. The revealing experience of this peculiarity never occurs in natural conditions; selection operates neither for nor against it.

These indisputable facts cannot easily be explained in terms of our present knowledge of genetics. We cannot be sure that the current concept of the genes applies to these modes of regeneration, nor to those reported in earlier chapters. The suggestion would seem to be that genes normally exist which are inoperative or "adaptive."

The information contained in a cell apparently exceeds that expressed by the phenotype. It is only revealed in exceptional circumstances such as regeneration, atavistic revivals, and embryonic regulation. Our exploration of genomes and collection of their informational content are very far from complete.*

The genetic potentialities certainly exceed those attributed by molecular biology to any given genotype. The genes play a determinant role but the orders they send out via the messenger RNA do not always produce the same results, for the materials available to the cell are not perfectly consistent.

GENESIS OF NEW SPECIES THROUGH RECOMBINATIONS OF GENES

New species have been obtained by upsetting the chromosomal set, either by hybridization or by treatments breaking down the chromosomes. Most such novelties occur in the plant kingdom; they are due to hybrids of different species or genera. The classic instance is the kohlrabi created by Karpechenko (1928). This hybrid of two crucifers (cabbage and radish) is sturdy; it has the large, smooth leaves of the cabbage but is sterile. It has 18 chromosomes, 9 from the cabbage and 9 from the radish, which, not being strictly uniform, do not pair off during meiosis; hence, the production of normal reproductive cells is not possible. But among them Karpechenko found one or two fertile plants that yielded

---

*Plant galls show how sensitive the cell and its genes are to external agents. The substances rejected by the gallicolous insect always induce the same reaction, ending up with a coherent structure bearing no relation to what the cells of the host plant would have made without it. The genes are unchanged but their products are metamorphosed. The cellular medium in which the cells evolve therefore has an importance that should not be underestimated.

seeds with $4n$ chromosomes and thus originated a systematic unit reproduced in *no* other case and constituting a new species.

The doubling of the number of chromosomes (called *allopolyploidy* in the case of hybrids) occurs accidentally before meiosis in the ovary and parent cells of the pollen or, alternatively, in the growing point of a bud. Allopolyploidy has been achieved after hybridization between species of tobacco, foxglove, *Crepis,* and plums. A few natural species (*Galeopsis tetrahit,* a labiate; *Spartina towsendii,* a grass; *Iris versicolor; Rosa willsonii; Aeculus carnea,* a horse chestnut, etc.) are spontaneous allopolyploids.

In animals, tetraploid or, more generally, polyploid breeds seem to originate accidentally during meiosis or a fusion of nuclei right at the start of embryogenesis. So far, we know of no authentically allopolyploid animal species.

Under heavy X-ray irradiation, the chromosomes of the germ cells of *Crepis tectorum* break down, and the pieces stick together again without mutual recognition. By selecting parents according to their chromosome sets, Gerassimova created a new species, *Crepis nova,* having the external appearance of *Crepis tectorum,* with which it is able to hybridize; the chromosomes of either parent do not pair off owing to their genic disparity.

Drosophila paulistorum is a tropical fruitfly that lives in Guatemala and southern Brazil. It includes six subspecies kept apart by ethological barriers. When they can be mated, it is found that the female hybrids are fertile whereas the males are not. These subspecies are, it should be noted, phenotypically indistinguishable; they are differentiated on the basis of chromosomal and ethological differences. A line of Colombian origin yielded, with another subspecies, fertile males as well as females. Dobzhansky and Pavlovsky (1971) effected a selection based on sexual behavior that produced a line whose offspring interbreed only with one another. This line corresponds to a nascent species, while remaining *identical* with *Drosophila paulistorum.*

These new species studied are the product of a combination of preexisting characteristics and not of creation. They do not involve creative evolution. They show that under certain conditions (compatibility of chromosomes with cytoplasm in parents of differing species), with hybridization followed by man-controlled selection, new combinations of genes can be achieved that correspond to systematic units of a lower order, subspecies and species.

The mechanisms described here are not operative in creative evolution. Evolutionary strains remain pure throughout history. Paleontology provides no examples of interphyletic hybridization.

IX

A New Interpretation of Evolutionary Phenomena

The biological sciences, like morphine, go to our heads before they let us down with a bump. They intoxicate, by affording the swift, ephemeral but exquisite illusion of understanding what we observe.

Now, understanding is as remote from observation as the Milky Way is from Earth. Between observation and understanding lies the whole vast expanse of conjecture.

Léon Daudet: *Writers and Artists,* Volume VII, p. 56, 1929

INTRODUCTION

Their success among certain biologists, philosophers, and sociologists notwithstanding, the explanatory doctrines of biological evolution do not stand up to an objective, in-depth criticism. They prove to be either in conflict with reality or else incapable of solving the major problems involved.

Nobody can be sure that evolution consists of acquiring characteristics by use or direct influence of the environment. Nobody can prove that phyla, classes, orders, and families have their origin in random mutations similar to those undergone, at all times and in all places, by living plants and animals. Nobody can assert that the organizational schemes are the work of natural selection.

Confining our attention exclusively, and for the sake of sound logic, to panchronic species, which abound and which for tens and even thousands of millions of years have been mutating without any noteworthy change, we should still be forced to deny any evolutionary value whatever to the mutations we observe in the existing fauna and flora.

202

After impartial investigation, which I have carried on for years, I am in a position to conclude that:

1. The Lamarckian and Darwinian theories, in whatever form, do not resolve the major evolutionary problem—that of the genesis of the main systematic units, the fundamental organizational schemes.

2. They fail to account for a great many fundamental aspects and phenomena of evolution.

3. We have not yet obtained from the fossil record all the information it is capable of yielding.

Freeing our minds of theoretical notions, wherever they may have come from, let us take an honest look at the phenomenon of evolution and, in all objectivity, set aside the accepted doctrines, notably every form of Darwinism.

I have provided that evolution is not

a random phenomenon

a continuous phenomenon

a phenomenon necessarily tied to any immediate necessity

the product of natural selection.

Moreover, it may be taken as proved that

since adaptation is seldom perfect, the living creature makes do with a compromise in respect to its environment (in the broadest sense); it survives, despite its comparative inadaptation, provided its physiological balance sheet is sound

interspecific competition is very far from being universal

death is more often blind and unselective than it is discriminating

It is *untrue* to claim that evolution, guided by natural selection, is always favorable to the species or lineage. It leaves in its aftermath a huge graveyard of failures and mistakes.

The "better adapted" no more displaces the "ill adapted" than the higher necessarily eliminates the lower. Panchronic species, as variable as those of more recent times, demonstrate that evolution and mutagenesis are two independent phenomena. The latter is continuous, the former is not.

In the main part of this book I have shown with a great many facts how far mutations fall short of the evolutionary variations that gave rise to phyla, classes, orders, etc. There is no need to go over this again.

In order to create, evolution demands new materials, such as genes formed *de novo*, or untried patterns of overprinted codons. It is not at all the same gene that, from one class of vertebrates to another, induces the tegumentary ectoblast and its mesenchymatous lining to form ganoid, placoid, or cycloid scales in fishes, epidermal osseous scales in reptiles, feathers in birds, hair in mammals. Every novelty demands its own genes, which are themselves also novelties.

No biologist worthy of his reputation can limit himself to criticism of accepted doctrine, however necessary and valuable it may be; he has to construct, and is able to do so if he can discard accepted ideas and view evolutionary phenomena from new angles in the light of recent advances in paleontology and molecular biology.

INTERNAL FACTORS IN EVOLUTION AND THE CREATIVE REACTIVITY OF LIVING THINGS

The true course of evolution is and can only be revealed by paleontology. The pioneers in this science, led by Cope, were ardent Lamarckians, but nowadays we realize that the environment does not have the direct influence that used to be attributed to it. Yet the living creature is not a passive object, it reacts and fashions itself according to the information reaching it, which it processes, as well as that containing within itself.

When we consider as a whole the evolution of a major systematic unit—class, order, or superfamily—certain phenomena become visible and force themselves upon us. In this respect the genesis of mammals is most instructive.

Comparing the different lines of theriodonts, we notice that every one of them has acquired mammalian characteristics, not always the same ones and not at the same points in time. These characteristics have been wrongly attributed to the outward signs of a *convergent* or *ecological* evolution which occurs when animals belonging to very different groups and unrelated to one another live in the same environment and adopt similar if not identical modes of living, acquire comparable organs, and similar habits. The transparency of pelagic animals, the dorsal fin of fishes, cetaceans, and ichthyosaurs, and the rows of spines (ctenidia) of insects harbored as parasites in the fur of mammals (fleas, Nycteribiidae, forficulids, Mallophaga, pupiparous flies, Coleoptera, etc.) are classic examples.

The appearance of mammalian characteristics in the various strains of theriodonts sharing the same evolutionary trend is not in the slightest a repetition of mutations occasionally observed in natural populations. For example, the macropterism attacking with varying rates of incidence most species of *Metrioptera* (tettigonoid orthopteran insects), the violet and reddish colors often reported in populations of crickets of *Stenobothrus* and related genera, are mutations modifying preexisting genes in the genotype of the various genera. They neither persist nor become integrated, in an evolutionary trend, as, indeed, no mutations do.

The phenomenon we have in mind is another one. The lines of theriodonts born of the same ancestral stock have three types of characteristics: those common to all, those specific to each, and mammalian characteristics unevenly distributed among them (see Fig. 3). For example, a mammalian characteristic A exists in three of them and is missing in another two. The tritylodonts and ictidosaurs are the two sublines richest in mammalian characteristics, but the other families of theriodonts, gorgonopsians, therocephalic bauriamorphs, and cynodonts are never entirely devoid of them. The cynodonts even appear to have been the privileged line from which the first mammals may have been born.

Hominization, operating through the four main lines of primates—tarsiers, lemurs, simians, and hominids—follows the same rules as "mammalization" among the lines of theriodonts. Achievement of elephantine and equine forms derive from the same process; we can say as much of many other zoological groups.

It is likely that, at the beginning, the lines of descent (which did not evolve synchronously) that formed high-ranking systematic units (class, order, superfamily) had access to roughly the same genome. According to neo-Darwinist doctrine, as they adopted separate ecological niches they did not retain the same mutations and diverged increasingly as time went on. But, if their separation became more and more pronounced, why is it that among theriodont reptiles all lines acquired mammalian characteristics, among Equoidea equine characteristics, among mastodons and elephants elephantine characteristics, among primates hominoid characteristics?

In most theriodont reptiles, the mammalian characteristics are very similar and affect the same organs, or practically so (bone of the mandible and of the skull; masticator muscles; sinews of these; differentiation, shape, and insertion of teeth). Suppose there are three mammalian characteristics unevenly shared among four lines of Theriodontia of the same stock S: A, formation of a secondary palate; B, lengthening of the dentary; C, displacement of the insertions of the masticator muscles.

A appears in lines 1 and 3 at different times, and is missing in 2
B appears in 2 and is missing in 3
A, B occur in 4, but B before A
C occurs in 3 and 4, and after A and B

First Possibility

The genes A, B, C corresponding to these characteristics exist in stock S, in which they are inhibited or latent; nothing discloses their presence.

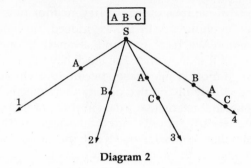

Diagram 2

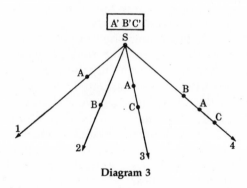

Diagram 3

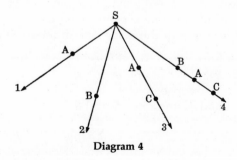

Diagram 4

They come into action at different times in the lines descending from S (see Diagram 2).

Second Possibility

The genes A, B, C are not present in stock S; they are present in it in other forms—A', B', C'—which, in the lines of descent, mutate at different times but identically, to become genes A, B, C (see Diagram 3).

Third Possibility

The mammalian genes do not exist in stock S; they appear at different times but identically in all four lines of descent (see Diagram 4).

After the first possibility, questions of the following type arise: Where do the mammalian genes A, B, and C come from? What causes inhibitions in S and activates them in the lines of descent?

If in stock S, genes A, B, C are identical to what they are in the lines where they are expressed, we are bound to conclude that *evolution takes place in advance in the ancestor*, without any external sign of its having done so—that it is inscribed in the DNA and remains there in a latent state, until such time as circumstances are favorable for it to be expressed in the phenotype of succeeding generations. Such *cryptic evolution*, owing to its wholly internal character, escapes natural selection unless the old idea of a germinal selection as first imagined by Roux and renewed by Weismann (*vide supra*) is resuscitated.

The second and third possibilities presuppose, without saying so, that evolutionary phenomena, mutations for the Darwinian school, occur in identical form in all lines of descent despite the low probability of their doing so. That several *adequate* mutations should occur in the same line is an amazing, almost unbelievable, stroke of fortune; but if repeated in all lines it ceases to be a stroke of fortune. Is a repetitive phenomenon a matter of chance? Is it not rather firmly established, a law? But in that case, what physicochemical agents dominate chance and cause genes A, B, and C to mutate indentically in lines often separated for tens of millions of years?

The truth is that contemporary Darwinism is in worse condition than appears at first sight. So far I have criticized it lightly by arguing as if the variations observed in fossils were mutations, but in fact they are *no such thing*. I have already stated my case; let us simply recall that these are not nondescript variations, but arise by successive additions in the same direction at an infinitely slower pace than the random modifications caused by mutation–selection in natural populations (see Ford, 1971), that the new characteristics are correlated with one another and with those preexisting. These go to make up the idiomorphon, the evolution-ary unit par excellence.

Not all the qualities or properties enumerated above are encountered in mutations. The evolutionary stages through which the theriodont reptiles passed to become mammals have no relation to disorderly and fortuitous changes. No paleontologist could maintain the contrary with-out renouncing his own science.

For Darwinian doctrine the test is unanswerable. Whichever possibil-ity is considered to be right, it cannot be accounted for, whether with evolution by internal factors or evolution by nonrandom variations. The theory is caught in a vise, whose jaws no dialectical subterfuge can force apart.

Each evolution that we know about in some detail (genesis of amphi-bians, reptiles, mammals, history of the various orders of mammals, and so on) forces us to admit that a phenomenon whose equivalent cannot be seen in the creatures living at the present time (either because it is not there, or because we are unable to see it) occurs in the course of it.

For this phenomenon the cell is both the instrument and the effector; it paves the way for the evolution of living things. It does so in accordance with the influence exerted on the organism by external factors and by certain internal ones connected with the chemistry of living things.

Recourse to internal factors to account for evolution has been at-tempted by numerous theorists, whether convinced Darwinians or more or less faithful disciples of Lamarck. Among the former, two names are outstanding, Roux (1881) and Weismann (1896, 1902). Staunch Darwi-nians, they enlarge the area in which competition (i.e., selection) takes place to include the molecule and the organs that constitute a living being.

According to Roux, the struggle is waged among the molecules, tis-sues, and organs; what is at stake is the intake of food which according to its quality, ensures greater or lesser growth, and consequently dif-ferentiation. To explain how the equids became adapted to running, it is rather rash to view it in terms of a pitched battle between digits, with victory falling to the third in every limb. Such an interpretation would appear to be hardly realistic and outright simplistic.

Weismann, the purest of the pure among Darwinians and a relentless foe of Lamarckianism, but a scientist of unquestionable honesty and objectivity, acknowledged and publicly denounced the weaknesses of the doctrine. For him, natural selection as conceived by Darwin, Wal-lace, and their contemporary disciples cannot explain the successive variations in a given sequence, the development of complex organs and correlations of parts, the regularity with which the useful variation turns up, the increase and decrease of organs.* Weismann imagines that natural selection operates at three levels: between individuals (Darwin and Wallace), between parts, organs, tissues, etc., of the same individual (Roux), and between germinal determinants (the equivalents of our genes) (Weismann).

The most important struggle is between the "vital, invisible" particles (genes), whether they reside in the body cells (soma) or the sex cells. These "determinants" are never identical with one another and their ability to "assimilate" food varies from one to another. The strongest take the most food, and their descendants prevail over the others. The result is that by the second generation the parts of the organism represented in the ovum by the strongest determinants are the most highly developed. This explains the dominance of certain characteristics; the determinants of organs favored by natural selection are better fed than the rest and are the winners in the competition, hence the accentuation of change in that direction.

What is the use of expounding the theory at greater length, when it is uncorroborated by any fact whatsoever? The discovery of independent genes, albeit included in the same thread of DNA, does nothing to support it. The idea of a struggle among genes cannot be tolerated in the current status of our knowledge. If there is a change in the genes, it is possibly the result of the direct action upon them of the cytoplasm, comparable to that revealed by experiments in the interbreeding of species belonging to different genera. Work by Godlewski (1906, 1911) (cross-breeding of Echinidae) and by Bataillon and Tchou-Su (1929) (tree frog × toad) has taught us that the chromosomes contributed by the spermatozoon of a species A are partly or wholly not tolerated by the cytoplasm of the ovum of a species B; they are altogether or partly eliminated from the achromatic nuclear spindle in the course of successive divisions. It sometimes happens that the chromosomes of the spermatozoon of a species C are rejected by the cytoplasm of a species D but that the reciprocal is not true; the chromosomes of the spermatozoon of D are tolerated by the cytoplasm of the ovum of C. All cases are possible, from total tolerance to total incompatibility between the female cytoplasm of a species and the chromosomes from the male of another.

This demonstrates the wide range of action of the cytoplasm on the chromosomes and, most likely, their most characteristic constituent, DNA.

However that may be, the existence of internal factors affecting evolu-

---

*Even if, according to Weismann, we invoke panmixia.

tion has to be accepted by any objective mind, but Weismann's hypothesis is unacceptable. We must look for something else.

Examining the latest writings on internal factors in evolution, we see that in fact what they are looking for is an antichance other than selection to explain the adaptive, finalized structure of living things. Spurway's (1949) writings are quite revealing in this respect. Struck by the fact that a group of related organisms is characterized by "possibilities of mutation" that determine its "evolutionary possibilities," she tentatively forms the hypothesis that a species, family, or class mutates more easily toward some genotypes than others—the spectrum of a group's mutation may determine its evolutionary possibilities. Her husband Haldane (1958) adopts Weismann's notion of an internal selection of genotypes and assumes that a "directional evolution" would not be due to changes in the environment and that any disturbance in an evolutionary process may, though harmless for the species, be lethal or sublethal for another related one, but would play a role in selection (which has long been accepted).

Researchers of high reputation, including Waddington (1957), Langridge (1958), Woodger (1959), and Whyte (1965), propose explanations of a similar nature, very close to the theories of Roux and Weismann but expressed in terms of genetics. Waddington writes cautiously and, concerning a few of the facts found in favor of germinal selection, says: "These examples are, however, not very convincing."

Whyte (1965) expounded at great length the position of the Darwinian reformers; he emphasized the importance of natural selection confined to populations and argued in favour of genetico-germinal selection, but all his arguments are purely theoretical. Besides, orthodox Darwinism emphatically rejects any intervention of internal selection; it accepts only selection operating on populations, or it condescends not to reject totally the idea of internal selection but goes on to claim that it cannot lead to anything important.

As a non-Darwinian, I am not directly concerned in this debate but simply note that biologists do, while remaining faithful to the principles laid down by the founder, recognize that these do not entirely account for evolution, and, in particular, that natural selection acting on populations is incapable of guiding evolution.

It is remarkable, to say the least, that none of the biologists or natural philosophers we have quoted as pointing out the inadequacies of Darwinian doctrine should have mentioned the real "evolution in action" of the fossil record. And yet it is this which compels us to credit internal factors. Consequently, I question the legitimacy of the dilemma into which some Darwinians seek to force us: either to accept that all evolution is the result of selection acting on random variations, or to invoke internal factors and thus stumble into an outdated and unfounded vitalism.

The dilemma is not a real one; it only exists in the imagination of the doctrinaires. At the risk of repeating myself, mutations do not explain either the nature or the temporal ordering of evolutionary facts; they do not account for innovations; the precise arrangements of the component parts of organs and the mutual adjustments of organs are beyond their capacity.

The intervention of internal factors in evolution is less "mystical" than that of chance or providential preadaptation. It fits perfectly well the Lamarckian theory. According to Wintrebert (1962), nothing is acquired by the living organism that is not the response of internal factors to an outside influence or ethological change. *Living means reacting, never undergoing.* Above all, it means not waiting around for a fortunate chance occurrence to save the situation.

I am not personally convinced that the tendencies toward the idiomorphon that are a fundamental characteristic of evolution most clearly demonstrated by paleontology are solely accounted for by the binomial "action of the environment/reaction of the organism." Evolution is readied and designed in the innermost parts of the living being and its cells in response to both internal and external influences. We gladly acknowledge the state of dependency between the creature and the environment but believe that the organism, itself processing its sensory, trophic, or other data received from without, is able to build something new. The means employed by the cell to create novelty escape us, but are nonetheless there.

What happened in the cells comprising the bodies of the Theriodontia which paved the way for the genesis of mammals? Probably reactions, molecular shifts depending upon the infrastructure and chemical composition of the cells.

Evolution is not the result of petty ailments, slight upsets of the living cell; it depends upon the physicochemical structure of the whole living creature and the properties determined by or emerging from it. The cellular composition of the Theriodontia was such as to imply the achievement of mammalian form. *Such genesis has nothing to do with chance, any more than it has with a vital principle.*

The paleontological facts referred to are not the only ones compelling us to look for causes other than random mutations and antichance selection; there are many, many others. When Étienne Wolff (1971) observes that in the little viviparous toad of Mount Nimba (*Nectophrynoides occidentalis*) there have developed in perfect correlation characteristics, barely hinted at and latent in other amphibians, which have made possible a strictly viviparous ontogeny, he raises the question of how such a complicated anatomical and physiological system, in a perfect state encountered only in mammals, can ever have formed in this tiny amphibian living in the isolation of damp mountaintop meadowland in Guinea.

The same—or almost the same—phenomenon is found in selachians (sharks), in which viviparity developed variously and in somewhat divergent directions. In these fishes, evolution favored a great profusion of inventions, such as the very long filaments, or trophonemata, on the inner wall of the uterus, which pass through the spiracle in tufts (first branchial postocular slit or vent) into the pharyngeal cavity of the embryo where they diffuse nutrients. Curiously, in the same genus *Mustelus*, in which all species are viviparous, only one, *M. canis*, forms a placenta.

Even odder is *Hemimerus talpoides*, a forficulid (earwig) that lives ectoparasitically in the fur of the Gambian rat (*Cricetomys gambianus*). This is a true viviparous animal, its tiny egg developing in the genital tracts of the mother; a placenta, formed jointly by embryo and mother, feeds the former at the expense of the latter. True forficulids all reproduce by laying eggs in pits in the ground and staying near and tending them. *Arixenia* of Java, which lives in the guano of bats, may be viviparous.

Were the genes controlling viviparity in *Nectophrynoides occidentalis* and placenta formation in *Mustelus canis* present in their direct ancestors, were they acquired in an accelerated evolution of those species alone, or are they latent, inactive, in other members of the genus? Here the problem joins that of cryptic evolution, except that in this case it cannot be doubted that the species are related or that they have most of their genes in common. This remark, however, is not necessarily applicable to genes specific to viviparity and placentation.

Some of the genes active in viviparity in *Nectophrynoides* exist in other amphibians, although they are but feeble in their effect: e.g., the lutein cells of oviparous animals are derived from the follicular cells of the ovary and contain traces of the hormone progesterone, but nothing else. In the viviparous selachians, which have been less studied than the Guinea toad, the novelties are, it would seem, more evident.

Granted that all the members of a group of oviparous animals possess genes capable of establishing viviparity, as a few species testify, it must be asked: Where do they come from? What causes inhibit or activate them? Who coordinates their effects so harmoniously? How would selection affect genes which do not perceptibly show their presence?

Étienne Wolff, wisely, estimates that we have too little knowledge to answer these questions now.

Possibly the Nectophrynoid reacted to an environmental influence, but it is to be noted that another amphibian inhabits the grasslands of Mount Nimba without having had to adopt viviparity.* All *Mustelus* species live in coastal waters and have no marked ecological or ethological peculiarities between one species and another; placentation in one of them is neither the result of selection nor of a reaction to the environment.

The *evolutionary work* of the living creature surprises us by its hidden causality and usually very apparent finality (which even those who purport to be antifinalists have in mind). But is it any more surprising than the adaptative reactions of unicellular animalculi to the circumstances in which they find themselves, some of the reactions being part of the heritage of the species and others not?

Among the latter organisms figures the resumption of normal position by *Arcella* after being turned on its back (*Arcella* is an ameba whose body is contained in a shell shaped like a beret, with cytoplasmic lobes or pseudopods projecting from it for use in crawling and taking nutriment); it sends out, as far as they can reach from the shell, long pseudopods which then contract and struggle to turn the shell over. But usually they cannot do so unaided, so a second reaction comes to the rescue: the cytoplasm secretes a bubble of gas that floats up to the top of the *Arcella* and displaces its center of gravity, whereupon the pseudopods are easily able to tip the shell over into its normal position with the opening toward the substratum.

So we have here the adaptive, nonhereditary reaction of an ameba— the simplest of all animals—displaying what the father of modern materialism, Ernst Haeckel, called the *psychological ability* of the cell (or the "lebendige Masse"). In a rich infusion of large amebas one may accidentally (animalculi have their accidents too) impale itself on a little thorn; it is thus captured, and likely to die. But what happens? The cytoplasm of

---

*With regard to reproduction, the amphibians are highly imaginative and adopt unexpected, even illogical solutions. In Amazonia, I have seen the enormous balls of foamy mucus in which the spawn of certain tree frogs (*Phyllomedusa*) is embedded and then dumped on the banks of permanent streamlets. Afer hatching, the tadpoles have to face what is for them a very severe ordeal in order to reach the stream in which they complete their development. Everything would be so much simpler if the spawn were dropped directly in the stream by the mother, as most tailless oviparous amphibians do. But what does a tree frog care about human logic?

the ameba splits (by invagination), or, better still, is cleft along a line parallel to the thorn (as a pat of butter is cut by a wire); when the split reaches the thorn, the ameba moves the two lobes apart and so finds itself liberated.

This fact, reported by Pénard (1938), the eminent Swiss protistologist and an observer not to be disbelieved however much it offends our sense of what is proper, shows us in a pure state the psychogenic and immediately adaptive ability of the cytoplasm. Anyone who has observed infusions has witnessed similar happenings; I have seen a big proteus-type ameba (*Chaos diffluens*) corner a ciliate infusorian in the angle formed by two filaments of *Spirogyra* and capture it in its pseudopodia.

The behavior of *Arcella* and *Amoeba* (I could write a short book about either of them) is not an *invention* of vitalist biologists. It has been observed by trained naturalists who know how to remain objective while facing the strangest spectacles that can be offered to their view by infusions or cultures, and able to measure the complexity of structures as well as the behavior involved.

I dedicate these remarks to those who would simplify the properties of living things to the point of insignificance. The observation of an animal in action in its proper environment remains an exercise as essential to the biologist as to the natural philosopher.

What the individual does, why should not a whole line do too? Cell reactions of a frankly individual nature become specific and hereditary; the proof is *Arcella* and its ability to right itself. It puts us right in the midst of reality.

The trends guiding evolution depend upon internal factors which we believe to be unrelated to the battle between parts imagined by Roux, and the germinal selection talked about by Weismann and the reformist Darwinians. These factors fit into a mechanism related to the physicochemical properties specific to every individual and its constituent parts. Paleontologists have observed that in a large number of lines, the size of the species increases as the group ages. Reptiles and mammals are excellent examples of this tendency. The first genera are of small size, even extremely small, whereas the last genera are the largest, sometimes becoming giants. Increasing or decreasing allometries are evident in the aging and in the growth of the line.

In my opinion, however, even more significant is the appearance of new characteristics, linked to a growth in size. The history of mastodontal mammals, rhinocerotids, dinoceratids, and reptiles such as ceratopsids provides us with examples of which the best is that of perissodactyl brontotherioids (titanotherioids)—best because of its many well-preserved fossils. The exhaustive work of Osborn (1929) allows us to state the phenomenon precisely.

The oldest of the brontotherioids date from the lower Eocene, the most recent and final representatives from the lower Oligocene, a history which lasts less than 30 million years. Their evolution is characterized by a continuous increase in the size of successive genera (from that of a large fox to that of an elephant), by a decrease in the relationship between the length of the face and skull, by a tendency to become tridactylous with the predominance of the third digit, a smaller foot, and a larger femur, and by the graviportal structure of the limbs.

Although we do not mean to attribute these tendencies to the increase in size, nonetheless, the production of horns seems possible only in the genera having surpassed a certain dimension. Horns first appear in the genus *Palaeosyops;* they are slightly longer in *Telematherium,* and are pronounced in *Manteoceras.* These horns are protuberances which grow above the eyes and form at the juncture of the nasal and frontal bones; they are thus nasofrontals.

"In the genera *Palaeosyops* and *Limnohyops* (middle Eocene) . . . most of the skulls were hornless, but some very old males of *Palaeosyops* show an incipient nasofrontal protuberance and roughening of the outer tabula of the bone" (Osborn, 1929, p. 266).

In *Manteoceras manteoceras* (middle Eocene) the protuberance, although small, is perfectly distinct and fully characteristic in form. With the genera *Megacerops* and *Brontotherium,* both of the lower Oligocene, the horns become very powerful.

The appearance and phyletic development of nasofrontal horns thus seem linked to increase in size and consequently to internal factors of a genetic nature.

It is doubtful that we have yet fully penetrated the complete significance of this phenomenon. Nevertheless, it is not presumptuous to hope that in the near future the mechanism of the genes which determine appearance and control size, whether in one species or in a whole line, will be elaborated.

Seen in this light, the credibility of the intervention of internal factors in evolution is at least as strong as that of a natural selection based on problematical variations.

THE CREATION OF THE NEW

Evolution does not confine itself to transforming what exists, whether it be enclosed in the body of a cell or recorded as a bit of information in the DNA strand. It creates, as we are bound to accept unless we wish to revert, in a roundabout and updated mode, to a preformist thesis resembling the "emboitement" of germs as formulated by Charles Bonnet in 1784 ("Contemplation de la Nature").

A mammalian hair requires information, and a corresponding executant, which the reptile lacks. In such case acquisition is certain. It seems to be the same for the breast. Reptiles are not totally devoid of integumentary glands, but these are sparse and, on the whole, undeveloped. The glandular cells of these vertebrates show, as Gabe and Saint-Girons demonstrated (1965, 1967, see also Chapter VI), a tendency to keratinize, whereas the integumentary glands of mammals accumulate fat and attain their perfect form in the sebaceous glands.

Thus, the creative and adaptive ability possessed by every living thing, however simple, has nothing to do with mystical properties, an entelechy in the sense of Driesch (1909) and Dawydoff (1942). It is found in the behavior of unicellular creatures, in plant "morphoses" (Bonnier's alpine plants, Molliard's anaerobic radishes). It is readily demonstrated by experiments.

This ability, which has nothing to do with instinct or intelligence, exceeds, in my opinion, the "cellular soul" imagined by Haeckel (1877); it is like the regulatory capability without which no autonomous natural system can subsist.

SIZE AND NEW CHARACTERISTICS

To speak of internal factors determining evolution immediately arouses the suspicion of many biologists. For them, it conjures up visions of the ghost of vitalism or of some mystical power which guides the destiny of living things by deciding their forms and regulating their functions.

My idea is quite different. Continuing in the line of Weismann, Roux, and many other biologists, I seek internal factors in the genome of the species, in its interactions with the surrounding nucleoplasm and cytoplasm. After all, isn't the gene the very epitome of an internal factor?

Moreover, I am familiar with certain evolutionary facts which imply the intervention of such factors. In recent years, zoologists have shown that there exists a connection between size and certain organs within a single species or a systematic unit.

Lameere (1915) was one of the earliest, if not the first, to study this phenomenon. The subject of his investigation was prionian beetles, several species of which exhibit an accentuated sexual dimorphism; the mandibles of the male are hypertrophied whereas those of the female remain normal. It is observed that "the secondary sexual characteristics are even more developed because of the prionians' large size, whether considered as individuals of the same species or by comparison with other species." In his largely forgotten book, Champy (1924) took great advantage of Lameere's discovery and amplified it. He was the first to make reference to the evolution of titanotheres and to the antagonistic growth which strikes certain organs and which is observed in one species or in one line, according to the individual's size.

In two independent studies of postembryonic growth, Teissier and Huxley (1936) discovered the phenomenon known as *allometry* and its mathematical formulation. Allometry is dependent on the expressivity of certain genes (internal factors *par excellence*) and on the cytoplasmic conditions in which they act.

No reptilian gland foreshadows the sudoriferous gland, merocrine or apocrine (holocrine). But the mammary glands of all mammals are similar to the sweat glands. Gegenbaur's (1886) ancient hypothesis that only the mammary glands of monotremes originated in the sudoriferous glands, whereas those of marsupials and eutherians came from the sebaceous glands, has been discarded in the face of the facts. But although resemblances favor filiation from the sweat glands, specifically mammalian, let it not be forgotten that mammary glands show a real independence even during ontogeny. Their earliest anlagen, the mammary ridges, ectodermal folds protruding into the subjacent dermis, begin at the axillae, symmetrically on each side, on the ventral surface of the embryo and end in the folds of the groin, whereas the sweat glands have no linear traces but burgeon *in situ* though growing only at a single point.

It may be worth adding that everything in anatomy and embryogeny testifies that the sweat glands and body hair have been closely related in their evolution. Do they not disappear almost entirely and at the same time in the same orders, Cetacea, Proboscidia, Sirenia, and Pholidota? Similar facts are observed whenever the specific characteristics of every phylum, class, and order are analyzed.

I am not alone in supporting the idea I launched, several years ago now, that creative evolution is not solely explained by the modification of preexisting genes but demands the genesis of new ones. This is what an American geneticist has to say on the subject: "Yet, being an effective policeman, natural selection is extremely conservative by nature. Had evolution been entirely dependent upon natural selection, from a bacterium only numerous forms of bacteria would have emerged. The creation of metazoans, vertebrates and finally mammals from unicellular organisms would have been quite impossible, for such big leaps in evolution required the creation of new gene loci with previously non-existent functions" (Ohno, 1970).

All this is rather obvious, but if people wilfully close their eyes to it, they will not see.

INDEPENDENCE AND PREEMINENCE OF DNA

Mutation implies a change, but not a creation. So we have to ask whether it is the only variation that may possibly have an evolutionary force. Many geneticists unhesitatingly answer in the affirmative, although evolution is not their subject.

Biochemists take as a pretext the heterogeneous structure of DNA and the transcription of its information by RNA to proclaim the dogma of DNA being not only the depository and sole distributor of the specific information available to the living creature, but of its presiding over the very genesis of that information. To chromosomal DNA has been added that of the mitochondria, plastids, and kinetoplasts, which has done nothing to abate the totipotence of this in so many ways exceptional substance. That DNA is the depository of the hereditary information is only indirectly related to the mechanism of evolution as we understand it.

DNA does not manifest its properties, let us say its powers, unless the cytoplasm (conceived in its totality) allows it to do so.

In the cytoplasm live the bearers of secondary information (hormones, pheromones, etc.) which specifically trigger the activities of genes. The fact cannot be contested, for it is easily observed. If the insect's body is deprived of ecdysone (the hormone of ecdysis), it is found that the giant chromosome of the salivary glands of *Chironomus* (a small nematoceran dipteran) larvae lose some of their DNA-rich puffs. At the same time the amounts of certain enzymes (hyaluronidase, protease) drop inside the cell. Some genes have been given a rest (Beermann and Clever, 1964).

The hormones are the greatest activators of DNA and the RNA's; their entire physiology shows it. But the prime information resides in, and emanates from, the DNA.

Let us consider the developing ovum. The information for the future organism is not only condensed into the chromosomal DNA; part of it resides in the cytoplasmic structures. The segmentation of the ovum has the effect of distributing the elements of the cytoplasmic information to separate sites where certain genes (chromosomal DNA, mitochondrial DNA) predominate. It creates local environments which stimulate certain genes and are conducive to their functioning; the activated genes modify the environment which, in turn, stimulates other genes and so on.

In the developing ovum, the embryonic organizer is a site at which materials having their own properties have formed during oogenesis and been installed during fertilization. In the constituent cells, certain genes are activated and trigger the generation of informational substances (chemically unspecified) which are diffused outside the organizer and set in motion the genes in the nuclei of certain blastomers. This is the property termed embryonic induction.

A steady flow of information leaves the nuclear DNA, while another flow comes toward it and triggers its activity. Mutual relations between cytoplasm and chromosomes (DNA), and vice versa, are constant and obligatory.

The flood of actions by genes depends upon triggers or chemical signals. In a blastomere A, a gene a induces formation of a substance S, which travels into a blastomere B where it triggers activity of a gene b and as a consequence formation of a substance S, which travels into a blastomere C, and so on. This outline is not very far from reality.

Even before recent discoveries weakened the dogma of the total independence of DNA and its preeminence in cellular functions and ontogeny, we were entitled to assume that the nucleic acids are not immune to enzyme or hormone action of cytoplasmic origin. There is no biological process involving a chain of chemical reactions that is not reversible, wholly or in part, and that does not determine some retroaction. The latter is by no means immediate, but often takes place discretely and with a time lag; although indirect, it is nonetheless real. And beyond the process, not always but very often, some kind of reserve system stands ready to moderate, accentuate, or even inhibit it. The regulator never loses its right to intervene.

Why should the nucleic acids have a privileged status in the living creature? It is hard to see. In fact, the partisans of a sovereign DNA have a mistaken idea of the hierarchy of the constituent cells and organs in living things. This forms a network of three-dimensional mesh, and not a pyramid with DNA at its apex. The links are not at all rectilinear and univocal, but are multivocal and often zigzag. This is equally true of the interior of the organism and of the effects produced in it by the messages received from the senses. The unity of the individual not only exceeds the sum of its juxtaposed and integrated parts, but imposes on them functional solidarity at all times and in all circumstances. Weismann's dualistic concept facilitates didactic exposition but dissimulates the underlying unity of the organism.

What happens in the fundamental cytoplasm remains obscure, though some of its component proteins are known. Its influence on the chromosomes has been demonstrated by several experiments conducted long ago and ignored by almost all contemporary biologists. Chambers (1917, 1921) priced with a fine stylet the first-order spermatocytes of the American cricket (*Dissosteira carolina*), and one minute later the chromosomes shortened and assumed the cruciform and hooked shapes characteristic of the pachytene stage and diakinesis. In another experiment the nuclei of spermatozoa penetrating into an immature sea urchin egg arrange themselves in unison with the nucleus of the ovum (Brachet, 1922); if the latter contains well-defined chromosomes, these also form in the male nucleus without going through the intermediate vesicular stages as during normal fertilization. As Brachet wrote: "The stage of the cytoplasm conditions the nuclear structure." Even earlier experiments with the centrifugation of *Ascaris* eggs by Boveri and Hogue (1909) have lost none of their topical value. They demonstrated that the fate of the chromosomes, including amputation of parts of them, is governed by the surrounding cytoplasm.

It must not be forgotten that the *germ* of certain animals *owes its properties* to the presence, in the blastomeres that will form it, of a particular cytoplasmic material, aptly called the *germinal determinant.* Thus the all-powerful germ finds itself during ontogeny dependent upon substances in the cytoplasm. The schematic action–reaction–action fits the case very nicely: In the ovum the genes activate the formation of the germinal determinant that will decide the fate of the blastomeres in which it is sequestrated by segmentation.

Through the life cycle from the ovum to the adult animal, via the genesis of the gametes and fertilization, DNA retains its structure: this is undeniable, but its activation depends on the circumambient cytoplasm. Ontogeny is only a sequence of actions and reactions between DNA and its environment. The organizers, secondary information centers, stir up certain genes while at the same time readying the materials they will need to use.

The organism is a whole. DNA alone can do nothing.

THE INTANGIBILITY OF DNA AND THE CENTRAL DOGMA

The dogma of the supremacy of DNA as sole custodian and always one-way distributor of biological information has been maintained by eminent biochemists (Watson, Crick) and geneticists (Jacob, Monod). This is what Monod (1971) wrote several years ago: "It is not observed, nor indeed is it conceivable, that information is ever transferred the other way round..." (pp. 124–125).

The ink with which this sentence was written was not yet dry when its flat denial was proclaimed. The logic of living things, which was, by the way, the biologist's and not nature's own, was overturned and the fine edifice cracked from bottom to top.

The discovery of enzymes capable of using viral RNA as a matrix for synthesizing DNA is regarded as a revolution in molecular biology.* As far back as 1964 Temin (1964) posed the hypothesis of the possibility of RNA forming a replica of DNA. Six years later, in collaboration with S. Mizutani, he showed that the RNA of the Rous virus, a pathogenic agent of fowl sarcoma, did engender by one of its enzymes, *reverse transcriptase*, a DNA replica which, in his opinion, would be incorporated into the chromosome of the host cell in the so-called provirus form. Under certain conditions the provirus DNA produces viral RNA, the initial model of the circuit: RNA→DNA→RNA. Independently of Temin, Baltimore (1970) arrived at the same result using Rauscher's leukemia virus in mice.

Shortly after this Spiegelman *et al.* (1970a,b,c) and his team announced the discovery, in seven varieties of oncornavirus, of a polymerase which synthesizes a complementary DNA of the viral RNA, thus confirming that the latter served as a model. Using the technique of hybridization of DNA with RNA, the same biochemists showed that the newly synthesized DNA did indeed mimic the viral RNA.

Hill and Hillova (1972) have observed that the isolated DNA of rat cells (XC) previously infected by the Rous sarcoma virus (Prague stock) transforms fowl fibroblasts, which start to produce infectious viral particles of the Rous sarcoma type. Confirmation of these facts spread rapidly. Todaro *et al.* (1971) demonstrated the existence of reverse transcriptase activity in two viruses with noncarcinogenic RNA (slow viruses causing disorders of the brain, notably in sheep). Thus, the enzyme is by no means tied to the carcinogenic property only. We have other proof of this.

Research by Hatanaka *et al.* (1971) furnished new facts in favor of Temin's observations. They introduced the mouse sarcoma or leukemia virus into a culture of mouse embryo cells (fibroblasts). Using autoradiography and tritiated thymidine, they found that a sudden powerful synthesis of DNA follows contamination of the cell by the oncor-

---

*And as the most important new discovery concerning the role of viruses in the genesis of cancers. Several viral RNA's producing DNA replicas are carcinogenic.

navirus DNA. In a permanently contaminated cell no such synthesis occurs, nor does it happen when the same viruses treated by ultraviolet rays are used. Hatanaka and his collaborators think their results are attributable to the fact that synthesis of the new DNA has been induced by the viral RNA. The DNA is not mitochondrial; it remains durably associated with the DNA of the host cell.

The work of Ross *et al.* (1972) and Kacian *et al.* (1972) carried out *in vitro* using a purified reverse transcriptase from an RNA virus (bird myeloblastome) and mRNA for synthesis of globin (regarded as one of the most typical messenger RNA's of mammals) has shown that globin mRNA with reverse transcriptase allows synthesis of a complementary DNA (gene of globin). Other equally important work shows that a reverse transcriptase acts in conjunction with RNA's not derived from the virus. I quote two examples having a direct bearing on our subject.

Beljanski *et al.* (1972) treated cultures of colon bacillus (*Escherichia coli*) with an antibiotic, showdomycin, isolated from a fungus (*Streptomyces showdoensis*). The whole population treated in this manner can very quickly become resistant and display new properties: synthesis of RNA noncomplementary to DNA, changes in structural and enzymatic proteins. Such observations are accounted for only by the hypothesis of a gene not usually expressed and "aroused" under particular conditions. These writers have demonstrated the existence of an RNA episome associated with DNA and made functional by the presence of showdomycin. Mechanically isolated and purified, this RNA has a "genetic" power of its own; introduced to a plant-tumifying bacillus (*Agrobacterium tumefaciens*), it modifies some of the latter's properties, suppressing in particular its carcinogenic ability. Bacteria transformed by the episomal RNA give off, as do showdomycin-resistant bacteria, an RNA capable of modifying other bacteria in turn.

The alteration thus obtained of the tumefying bacillus is hereditary, so it must have affected the genetic code of the microbe.

Beljansky suggests an explanation: The RNA introduced into the acceptor bacterium is interpolated between certain RNA's already associated with DNA (e.g., by means of ligases), thus creating a new RNA episome capable of imposing its own information on that of the host. He and his collaborators have proved the existence in bacteria of a reverse transcriptase able to synthesize a complementary DNA precisely modeled on the transforming RNA's, episomal or excreted RNA, both having the same genetic potential (Beljansky and Manigault, 1972).

Tatum and his collaborators (1975) presented evidence that RNA of wild-type *Neurospora crassa* was capable of inducing inherited transformation of inositol independence in *Neurospora crassa*. This fact shows that RNA completely free of DNA can be used as a vector of genetic information.

The work mentioned proves the existence of a molecular mechanism which in given circumstances brings the organism external information and fits it into the DNA of the genetic code. For an evolutionist, this is of tremendous importance.*

ACQUISITION OF THE INFORMATION BY THE ORGANISM

Information forms and animates the living organism. Evolution is, in the end, the process by which the creature modifies its information and acquires other information.

Extrinsic or intrinsic factors act upon the DNA through the intermediary of enzymes or RNA. Nothing evolutionary can be inscribed in the genetic code without preliminary work by the entire cell. Mutation is an accident or disease having only a remote bearing on the evolutionary process; this is proved by the independence of mutagenesis with respect to evolution.

Inscription of evolutionary or adaptive reactions in inheritance requires special conditions. Now, we know, and must bear in mind, that as the world of living beings has grown older, evolution has never ceased to dwindle. Why are evolutionary reactions becoming rarer? In our present state of knowledge, it is futile to ask. When molecular biology has increased in accuracy and refinement we may be able to find the answer.

And yet we already have at our disposal facts which admittedly provide no solution to the problem of evolution but do help us to understand its phenomena better and to orient our research along paths not previously taken.

The animal would not survive if it were not informed about its environment in the broadest sense of the term. The sense organs receive messages and transmit them in modified form to the nerve centers, which process them and elaborate adequate answers to the external stimuli. The nerve centers, the organism's variable input computers, are themselves the product of specific inborn information and subject to permanent monitoring by it.

The specific information resides inside each cell in the strands of DNA

---

*The phenomena of viral replication, transcription by DNA of information of RNA origin, and many others are only partly explained. We should not have too many illusions about the value of the interpretations offered.

and is identified by the genetic code.* It is the intelligence of the species, in highly miniaturized form. It is also the mind of the lineage at a moment of evolution. It has been established, integrated, and inscribed itself in the DNA during the stages passed through by successive species. It is the end product of a slow elaboration, as a state of equilibrium was gradually arrived at between the creature and its environment.

Specific information is transmitted in the form of chemical signals sent out by segments or genes of DNA. Minute analysis reveals that it is contained in this molecule in the form of codons (sequences of three consecutive nucleotides forming the code of an amino acid) comparable to bits of information; the gene may be understood as constituting a set of codons.

But according to Darwinian doctrine and Crick's central dogma, DNA is not only the depository and distributor of the information but its *sole creator*. I do not believe this to be true.

Left to itself, DNA undergoes, during its replications in the germinal cells, the mutations so often referred to in the body of this book. But error modifies what already exists, it does not create it.

A library does not fabricate information, it receives it from without, classifies and stores it. The medieval copyists made mistakes that altered, vitiated the texts they were supposed to reproduce. Who dares assert that their errors are the work itself?

If Crick's thesis is correct, the first living creature, as we have said, already had in itself *all* the genes which have enabled the genesis of plants and animals in all their infinite variety. Since nothing could have come from without (as postulated by Weismann and updated by Crick) except the disturbance caused by mutations, DNA has necessarily to be the only generator of the biological information. To attribute such a power to a single substance, however complicated and exceptional its molecular structure may be, is in my view aberrant. Actually, to compel the living creature to evolve, DNA has to receive messages either from other parts of the cell or from organs (e.g., hormones) or from the outside world (sense stimuli, pheromones, etc.). Of itself, by what miracle could it generate information adequate to performance of a given function?

---

*The information in an organized being is not at all localized as in a computer. Each cell, i.e., each executant, contains the whole of it. There are as many "memories" as there are cells. The chief function of the nervous system is to process information from the outside world. Nevertheless, the genetic code is partly under its direct control. The purpose of these remarks is to show the gulf separating a living creature from any man-made electronic appliance, computer, or robot. The cybernetic lineaments of the living and the nonliving are not the same. Analogies between them concern parts and not the whole.

The whole range of mutations, or *mutational spectrum*, of a species has nothing to do with evolution. The "jordanons" (mutants) of whitlow grass (*Erophila verna*), the wild pansy (*Viola tricolor*),* plantains (*Plantago*), candytuft (*Iberis*), which constitute well-catalogued and rich collections, are irrefutable proof of this: they are not derived from one another, and are indefinitely stable. They display the species with all its collection of invariant variants, translating, so to speak, oscillations in the polymorphy of a specific unit about the equilibrium state of a genome in its environment. Thus, despite their innumerable mutations, *Erophila verna*, *Viola tricolor*, and the rest do not evolve. *This is a fact.*

The catalogue of breeds of dogs, as of any other domestic animal, is simply the mutational spectrum of the species, sifted by artificial selection. The same can be said of the list of varieties of any cultivated plant. Nothing of all this constitutes evolution.

Any acquisition of new information requires a structural change, something added. It is not at all a matter of altering or suppressing one or more preexisting units, but of *adding more.*

The computer is limited in its operations by the program controlling it and the units of information fed into it. To enlarge its possibilities, its contents have to be enriched. What is new comes from outside.

In biology argument by analogy endangers interpretation, for it applies our Aristotelian logic to living phenomena; as I have not failed to point out in earlier chapters, cybernetics teaches us what is by no means a negligible fact—that certain control models have to be used both for the natural self-regulating systems formed by living things and for inanimate physical or physicochemical systems. The cybernetic model, of which philosophy has not yet fully taken advantage, is applicable to all kinds of biological systems, whether relating to structure or to function—both being closely interdependent.

An engineer who knows no biology often builds a machine based on principles already applied in living things. Such convergence indicates that there are not so many different routes to reach a given end. Sometimes there is only one.

---

*Darwin (1868) failed to recognize the constancy of the jordanons of *Viola*. Wittrock (1903), who took up the study of *Viola tricolor*, confirmed the "absolute fixity of these types of *Viola*." Naudin, who cultivated the "elementary species" of *Erophila verna*, found that after 10 years the plant had subdivided into 10 elements, after 20 years into over 50, and after 30 into over 200. "It is undeniable," he adds, "that a good many, if not all, Linnaean species are simple collections, often numerous, of refined forms. Jordan's cultures prove that elementary species do exist and remain autonomous, without merging into or hybridizing with one another, as all good species should...." It seems to me that Naudin's "refined forms" and "elementary species" are indeed mutants.

To my mind, the comparison between certain functions of the living creature and the operations of a computer has no reductionist implications. It simply tends to show that both the machine and the living being have to be programmed and fed with external information in order for novelties to emerge.

But let us return to where we were a few pages back. Certainly the information in DNA, just like that contained in a book, is released and activated only by external intervention; its bearer, specifically stimulated, transmits messages in the form of sequences of RNA (mRNA) and reproducing the arrangement of the nucleotides of one or more genes. Ordinarily the mRNA travels from the cell nucleus to the cytoplasm.

To be understood, the message of the mRNA has to find in the cell a *receptor* capable of decoding and deciphering it. The receptor consists of ribosomes rich in RNA, amino acids, and sundry compounds; through it the induced message directs the synthesis of a protein (holoprotein, or heteroprotein with a prosthetic group), but an enzyme, RNA aminoacyl synthetase, has to be present to catalyze the reaction. Now, the biological origin of this enzyme apparently affects the order of amino acids in the synthesized protein. May we not infer, so far as results of *in vitro* experiments can be extrapolated, that the synthetase also acts as an informer in protein synthesis?

The state of the cellular medium in which the reactions take place is the result of synthesis; among active factors are cited pH, temperature, concentration of magnesium ions; "initiation factors" of a protein nature are also mentioned: three protein factors, IF-1, IF-2, and IF-3, are said to be necessary for fixing the triggering tRNA to the messenger for it to be linked with the ribosomes (Sabol *et al.*, 1968, 1970).

External factors reach the genes in various ways. For example, stimulation of the sensory receptors triggers in certain cerebral neurons a neurosecretion whose products travel along the axons to the endocrine glands. There they cause secretion of hormones which on being released in the organism activate the genes of *certain* cells.

Let us quote as examples of the long chain actions and reactions, starting in the case of vertebrates from visual stimuli (patterned or purely luminous) and reaching the brain (hypothalamus, hypophysis) and the endocrine glands (thyroid, suprarenal, testes, ovaries) that secrete gene-activating hormones. In the case of insects too, the chain reaction has a sensory origin (visual, antennal) but passes through the *pars intercerebralis* (or another zone of the brain) to the endocrine glands (corpus allatum, ventral gland) which release the gene-activating hormones. Such activation is shown by the formation of puffs at certain loci of the giant chromosomes of the salivary glands of Diptera and is triggered by ecdysone. Even in the ovum DNA can do nothing when unaided. The enzymes brought by the spermatozoon activate the ovum and modify its infrastructure and architecture. It is after an upheaval of the cytoplasm that a few genes come into play and launch synthesis of the enzymatic proteins which govern the differentiation of the blastomeres and embryonic layers.*

REPETITION OF THE SAME GENE, AND REDUNDANCY

Genetics textbooks are extremely discrete about the formation of new genes. They ignore this problem, of primordial importance in any explanation of evolution.

According to Watson and Crick (1953), the double helix DNA molecule replicates by opening longitudinally like a zipper; each half acts as a template for a new complementary half that remains attached to it and builds itself out of the nucleotides in the nuclear sap. This accounts for the reproduction of genes, which can occur only lengthwise in the DNA ribbon where the genes are lined up side by side in single file, like the beads of a necklace (Morgan, 1911).

The discovery of repeated, or redundant, genes inside the same chromosome prompted the notion of genesis by interpolation. But the interpolation of nucleotides or codons into a strand of DNA has been reported only in the integration of a virus (prophage) in the host DNA and viral transduction (discussed later in this chapter) of fragments of DNA from a first host cell. But such processes have only been able to create a sequence in which the same gene is rejected several hundred thousand times.

The conversion of a very long sequence of codons distributed in a disorderly way, to a series of short sequences (genes) of identical makeup, is a possibility that is not absurd, but is purely hypothetical.

---

*All ontogenesis is the realization of information which sends its messages according to a very precise timetable. From the organizer comes a substance which, as it spreads, creates a gradient of concentration, or field gradient, whose value diminishes with distance from the transmitting source. This interpretation is based on the work by Spemann *et al.* (1938). Other hypotheses are suggested, but are less plausible and do not have the experimental backing of the first. The Goodwin–Cohen model relies on a comparison of a source periodically emitting chemical signals with a pacemaker for the heart; this is a hypothesis based on migrations and agglomerations of mycomycete amebae—a transmitter ameba sending out rhythmic puffs of certain substances answered, by positive chemotaxis, by the receptor amebae; but in this case it is not a matter of cellular differentiation.

And we also ought to ask a prior question about the origin of the lengthy sequence.

The adding on of nucleotides to the free ends of the strand of DNA seems more probable. This is the way in which synthetic (or subsynthetic) nucleic acids prolong their molecules. However, in bacteria and peridinians, the only creatures whose form of chromosomal DNA is known, the fiber is closed upon itself in a circle. But if the circle is opened, the addition of nucleotides may become terminal and closure takes place later, once the addition is complete (see Chapter VIII and later in this chapter for what is said about chromosome breaks and joins).

It is pointed out that redundant genes have been observed in various regions, not necessarily terminal, of the chromosomes. It is supposed that iterative genes control the quantitative characteristics polyallelically determined, but this has still to be proved.

Mutation of the gene is probably increased by redundancy without any greater number of types. Without precise figures, the assertion is only a hypothetical one.

FORMATION OF NEW GENES AND PROBLEMS IT RAISES

No formation of new genes has been observed by any biologist, yet without it evolution becomes inexplicable; I have given many reasons why.

Molecular biology supplies the evolutionist with the means of conceiving or predicting the conditions under which such a synthesis would be possible. What I propose is not a gratuitous supposition but a deduction from well-established facts about the structure and synthesis of nucleic acids and the enzymes controlling them.

Remember, the strand of RNA synthesized by the half-molecule of DNA is a single helix which is detached from the template in which it is molded and drops into the nuclear sap. Conversely, the freshly formed DNA stays attached to the half-molecule of which it is the exact complement.

These are the minimum required conditions for formation of a new functional gene:

1. Addition of codons to the pieces of chromsomal DNA
2. Formation of an enzyme which activates the new gene by opening out the DNA molecule at the required level, thus enabling a messenger RNA to form

3. The enzyme should only be produced in the cells in which the gene will perform a precise action in agreement with, or at any rate complementarily to, the other functions of the living things.

To be functional, the new sequence of codons should make biological sense, i.e., transmit an mRNA carrying a message triggering synthesis of a protein performing a new function or perfecting an existing one.

Let us take a closer look at this. If the new gene is formed by adding nucleotides at the free ends of the DNA molecule, there is no template (half-molecule). The new sequence has to be a single helix. But according to Temin's findings, it can be duplicated by the action of an enzyme. Temin and his collaborators discovered an enzyme activity capable of altering the piece of single-helix DNA to a double helix at the expense of the virus RNA/host DNA hybrid. Another enzyme chops up the DNA strand into pieces, a property unrelated to our present purpose.

Does the DNA frequently forming at the end of a double helix molecule have any gene activity? Nobody knows. In the absence of a template which transmits significant messages, the added nucleotides are of no value except by the order in which they come; theoretically there is nothing to prevent them from organizing into codons which then order the formation of a protein used in the economy of the living organism.

If the synthesis of the gene is limited to the above operations, the biological significance of the new segment of DNA is a matter of pure chance, because the nucleotides are joined in random order. The newly formed sequence of codons might fortuitously assume a biological meaning, just as Émile Borel's monkeys using a typewriter might eventually make sense. The cell would be no more inventive than the unknowing animals.

Theoretically, a gene thus formed can function only with the aid of a specific enzyme that opens out the DNA molecule at its level and enables synthesis of the message-bearing RNA.* Such a requirement di-

*Rudkin's work (see Beermann, Chapter VIII) on replication of the salivary chromosomes of Diptera strongly supports my interpretation. He says that the chromosomes are not replicated *en bloc* but by sectors, constituting so many *units of replication* numbering several thousand. To each unit (cistron) would correspond a particular polymerase. But it is not sure that this enzyme would be the only one to interfere. It is not the same as the one causing, in all forms of mitosis, including endomitosis, all the molecules of chromosomal DNA to open and the number of chromosomes to be doubled (at the level of each half-DNA molecule there forms during mitosis the complementary half-molecule of the same nucleic acid). The elective activity of the cistrons has been verified *de visu*, not only in the giant chromosomes of Diptera but on the fiber of colon bacillus DNA (Miller *et al.*, 1970).

minishes the chances of a random successful synthesis since formation of the enzyme is just as unlikely as that of the gene. In order to create, evolution has to win not just on one count, but on two or even three. In theory this is possible, but a low probability is not far from zero. Besides, is it not presumptuous to try to explain a phenomenon that has held to a precisely plotted course for thousands of years, by a mechanism based on the most slender expectation of success?

Mindful of the fact that the genesis of mammals and their orders has been a slow and steady climb toward strongly marked idiomorphons, I reject the overly easy explanation of the random acquisition of new genes. The biochemical and molecular mechanism set in motion by the *joint* formation of a gene and the enzyme that will unlock the corresponding DNA segment cannot be aleatory in its essence. It is likely that one of the most potent internal factors in evolution concerns the formation of new genes and its control system.

Orientation and control are observed in every embryogeny, and no one would deny it. In the case of a phylogeny in which the orientation of phenomena is as plain as in the development of an egg, controversy arises, and some biologists are scandalized by the idea that evolution follows an appointed track. Is it so very difficult to stick to the facts?

The formation of the gene is profoundly different from the mutagenic process altering the order of sequence of the nucleotides that form the codons in a segment of DNA. To my knowledge, nothing else is involved in mutation, whether in the enzymes actuating the gene or in the cells in which it will operate; the system remains ready to function.

The machinery of creation is much more complicated than a disorder of the genetic code. It involves several constituents of the cell and directly depends on the latter's state, which is subject to external influences.

As stated earlier, the internal factors in evolution, the reality of which is established by the study of the lines of fossil animals, are probably partly the same as the mechanisms that form genes and their satellites or helpmates, the enzymes.

The difficulty is to catch these mechanisms in action while evolution slumbers and mutagenesis apparently carries on unchanged. But it may be that in certain living things biologists arouse evolutionary potentialities that one is tempted to believe are already exhausted. To make no

---

Ribosomal RNA is simultaneously synthesized by the cistrons 16 S and 23 S, thanks to a fairly large number of DNA polymerase molecules; the two cistrons occupy sites remote from one another.

distinction between the mutation and the creation of a gene is truly to understand nothing of the innermost mechanism of evolution.

The internal selection described by Weismann and his latter-day disciples, the Darwinian revisionists, must, if it exists, operate on new genes and enzymes; it is hard to visualize what means are used to get there. If the new gene is alone, without its actuator, i.e., nonoperational, it is unlikely to disturb the life of the cell; it is tolerated, and it continues unaltered. The DNA segments reputed to be genetically inactive and whose mass is, according to the classic writers on molecular biology, considerable, may be genes whose formation has not been linked with that of the actuating enzymes. In that case, whether they have any meaning or not, they lack the ability to express it; they have only the enzymes which ensure replication of the DNA.

Selection does not have anything to do with unfinished novelties: How could it act upon a function still to come? It only goes into action, if at all, when the gene is actually functional and affects the physiological equilibrium of its bearer.*

A New Concept of the Gene and Gene Overprinting

The notion of the gene was implicit in particular theories of heredity long before Johansen (1911) coined the term. Naegeli's (1884) idioplasm, Weismann's (1892) determinants designate roughly what we call genes. Observations and experiments have constantaly confirmed and corroborated this notion, which has made an immense contribution to the advance of our knowledge of heredity.

The DNA molecule, whose chemical structure was revealed by Chargaff (1950, 1965) and whose configuration was made known by Watson and Crick (1953), does not contain in itself any isolated particle; it is an edifice whose continuity is ensured by the hydrogen bonds between the

---

*Mills *et al.* (1973), using a nucleotide sequence of an RNA from the bacteriophage $\phi\alpha$, a parasite of the colon bacillus capable of reproducing itself *in vitro,* have attempted to trace the possible precellular evolution (*sic*—prebiotic would be a more suitable word for the experimental data in question). They report instructive facts about the reactions of certain RNA sequences to various substances: elongation, second or third order shortening, but of no obvious relevance to evolution. Going much farther, Mills and his collaborators imagine selective factors acting on the various sorts of RNA strands. Biologically, I do not see how in an inert physicochemical system factors favorable to a future evolution can be discovered, and the operative modes of selection deduced from them. From this study, I note that the RNA sequences modify themselves according to the environment in which they are placed, which is not by any means devoid of interest. But Mills *et al.* have *no* assurance that their experiments in any way recapitulate a possible evolutionary phase of the "prebionts."

complementary nucleotides. *Hence the gene can only be one segment of the molecule, whose site is revealed by recombinant genes* resulting from cross-overs of the chromosomes. We draw what we call the genetic map of the species by plotting on straight lines that represent the chromosomes the points or segments corresponding to the loci of genes (some, of course, not all) of the creature concerned. These are the A B C's of genetics.

At the present time it is accepted that the gene's function is to induce synthesis of a specific protein, nearly always an enzyme, hence Beadle and Tatum's aphorism (1941): "One gene, one enzyme." The gene is the DNA segment codifying the information required for building this protein. When it goes into action, the molecule opens out at that point and provides a model for a messenger RNA in which the complementary ribonucleotides are repeated in the same order as the deoxyribonucleotides in the DNA* and which remains in a single-stranded state. As we have already said, the messenger RNA comes out of the nucleus and gives its message to organelles, the ribosomes, themselves made of the particular RNA and ribosomal proteins, using the amino acids contained in the cytoplasm.

The code enabling transcription and transmission of the message is made up of units of information, or *codons*, made of three nucleotides.† Without going into all the detailed hypotheses and interpretations which the reader will find in the classic texts on genetics, let us note that a gene is a succession of nucleotides, every three contiguous ones forming a codon; each codon carries the necessary and sufficient information for determining the position of an amino acid in a protein being synthesized.

The ribosomes, granules grouped at particular points in the cytoplasm, are able to decode the message, read it, and, with the aid of enzymes, synthesize the proteins in the order received.

It is not enough to imagine the manner in which a new gene is born; we also have to know, and what it is important for the evolutionist to know, what limits it spatially and makes of it a functional unit. Mendel's laws of heredity and the facts revealed by molecular biology testify that a characteristic is a whole that remains identical with itself through succeeding generations. Therefore, it cannot be determined by a variable cause, and the gene controlling it has to be of constant composition to ensure that its effects are constant. Moreover, if the genes were not

---

*The nucleic acid is DNA, when DNA replicates itself.

†There are four kinds of nucleotides, depending on the bases they contain (adenine, cytosine, thymine, guanine). In RNA the thymine may be replaced by uracil.

circumscribed exactly, it is hard to see how their messages could be deciphered.

The consequences of chromosome crossovers, shifts, and inversions are such that we are entitled to conclude that the breaks pass between the genes. It is a disconcerting fact, but there it is. An experiment proving the reality of it was carried out by the Russian scientist Gerassimova.

The germinal cells of *Crepis tectorum,* a plant of the family Compositae, were heavily irradiated by X-rays. The chromosomes fragmented, and their fragments were stuck together again, end to end. By selecting the reproducers, Gerassimova obtained individuals with the same number of chromosomes as their parents but having changed places with one another. These individuals are sexually isolated because, since their chomosomes are incapable of pairing during meiosis with those of their parents, hybridization is impossible. Hence *Crepis* with a rearranged chromosome set forms a new species, *Crepis nova,* in every respect resembling its parents, *C. tectorum.* The fracturing and subsequent recombining of chromosomes notwithstanding, the messages sent by DNA are perfectly decipherable and produce the same effects. Logically, we should conclude that all the breaks passed between the genes.

In some cases the genes are not separated by silent points. Studying the locus *rII* of bacteriophage T4, Benzer (1959, 1961) found that the action of this locus is due to two adjacent functional units of complementary effect called *cistrons;** there is no indication that the two components are separated by a space equal to a punctuation mark.

The classics of genetics accept that a cistron is composed of one or more genes occupying separate sites (termed mutational because they are revealed by a mutation) and acting synergistically and complementarily. The location of the genes of the cistron is determined by the usual technique of recombinations of characteristics consecutive to chromosome crossing over; it has been recognized, in the case of *Drosophila,* for the lozenge-shaped eye characteristic, and in the fungus genus *Ascobolus* for a chemical property.

I shall not go so far as to claim, like Tavlitzki (1972), that "the term gene has no longer any meaning" when genetic analysis "is refined"; for the determination of all hereditary characteristics is still connected to one or more DNA segments. Nor does it matter what name is given to these.

---

*The cistron is defined as the smallest functional genetic unit.

Let us classify and generalize what we have written about the developing ovum and the actuation of genes.

A segment of DNA only opens out and forms a message if the cytoplasm and the nuclear sap are in a given state—probably if they contain an activating agency, enzyme or hormone. The cell goes through constitutive and physiological states that succeed one another in a fixed chronological order. The states, needless to say, vary from one organ and one category of cell to another. During each phase *different genes become stimulated*.

Take the genes D, M, R, S, W of an insect. D goes into action in the blastodisc of the embryo; M, much later in the differentiation of the nervous system; R only in the larva's imaginal discs at the start of metamorphosis; S at the end of metamorphosis during genesis of the scales; W for synthesis of a pheromone inciting the female to copulate.

No doubt, as we knew long before Gurdon's experiments (1963), all the cell nuclei of a metazoan contain the *complete* genetic code for the species; but it only issues its commands when excited by the environment, i.e., the cytoplasm around the nucleus and the nuclear sap in which the chromosomes are immersed. Examples have been given in earlier chapters.

Commoner [(1968) and various other publications] has proved by sound arguments that during the synthesis of specific proteins various biochemical factors (notably polymerases and synthetases) supply information complementary to that of the DNA.

The actions of enzymes and hormones play a predominant part in the life of the cell, in which effectors and reactors continually influence one another. It is found that DNA of itself does not initiate actions, but operates only when moved by certain principles. There is no need for a minute analysis of proteinogenesis in order to realize that independence is a vain word for DNA. Without the specific action upon it of enzymes and hormones, it is as inert as a quartz crystal. We are now able to say that DNA has a constant structure which only mutations disturb, and its activation depends on substances external to it. Moreover, the various activities of the cell take place in a fixed chronological order not directly controlled by DNA: an extrachromosomal "enzyme clock" triggers and times the entrances and exits of genes.

Supported by these facts, I conceive of the genetic code as something other than a row of genes, and propose an interpretation which, without rejecting the classic theory, provides the key to problems not yet solved. There is nothing to prevent one and the same locus (or one believed by us to be the same) from functioning more than once in the life of an organism, and in a different way. Let me explain: Take three adjacent

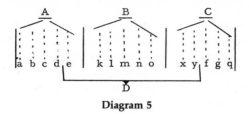

Diagram 5

genes *A*, *B*, *C* belonging to the genotype of an insect, of which we shall for the sake of clarity deliberately indicate only a few codons (see Diagram 5).

> *A* functions in embryogeny
> *B* functions in metamorphosis
> *C* functions in the mating of the sexes

Outside these periods, it is accepted that the long segment *A*, *B*, *C* remains at rest. But is it not active in a different way and at a different time from the three recognized and separate genes? We are entitled to ask the question. What counts in the gene is the order, number, and quality (according to their base) of the deoxyribonucleotides, i.e., codons. Let us assume that in the trio *A*, *B*, *C*, under the effect of a substance *S*, the codons combine to form a fourth set, or gene *D*, comprising the codons *e*, *k*, *l*, *m*, *o*, *x*, *y*. It may be that such a group forms a cistron and controls a different characteristic from those determined by genes *A*, *B*, and *C*.

There is nothing in the teachings of genetics or molecular biology to oppose this hypothesis, which assigns to DNA wider capabilities for storing information than the classic theory allows.* It explains that the

---

*Since the publication of the French edition of this book in 1973, the basis for the hypothesis of overlapping genes has been demonstrated by the discovery made by three British biologists, which they announced as follows:

"Bacteriophage φ X 174 genes *D* and *E* are translated from the same DNA sequence but in different reading frames" (Barrell *et al.*, 1976, pp. 34–41).

The genetic code is read, then, not exclusively according to the linear order of the genes, but by the overlapping of two or more genes or else by limitation to a segment within a gene. According to the classic thesis, the genetic code acts like the perforated roll one puts in a player piano which permits the execution of one piece of music only. According to my concept, which is confirmed by the discovery of Barrell *et al.* (1976), the genetic code can be read in various ways; it is comparable to the keyboard of a piano from which the performer draws an infinite number of musical pieces. The great problem is to discover the source and the nature of the information which recognizes and animates the genetic segments of the DNA.

amount of DNA does not necessarily correspond to the number of genes. Thanks to this hypothesis, of *overprinting*, we can understand that the amount of DNA may paradoxically be greater in a lower verte- brate than in a mammal whose structure and functions demand a higher number of genes than are required by fishes, amphibians, and reptiles.

There appear to be two preconditions necessary for overprinted genes to be able to function:

1. The genes encroaching on one another come into action not simultaneously but consecutively, according to a fixed program

2. The limits of the genes or cistrons must not impede the recombi- nation of codons

"Punctuation" genes, if there is such a thing, can only be a sequence of codons whose functions are not clearly apparent (spaces?). In fact, the DNA molecule is like a keyboard played by the chemical stimuli that determine the limits of the gene and render it operative, i.e., transmit- ting an RNA-encoded meaningful message.

What substances are responsible for activation of genes and for their length? We know the hormones (ecdysone in insects, estrogenic hor- mones in vertebrates) whose specific action bears on a strictly limited sequence of codons. The hormones themselves are the indirect product of an operation in which DNA participates by initially encoding the proteins which, reacting with ions, metabolites, vitamins, etc., in the cytoplasm, synthesize the hormones or differentiating principles (or- ganizers in the embryo).

The release of an enzyme (by breaking open lysosomes?) enables rep- lication by remote control of complex organelles such as the parabasal body (Golgi apparatus, flagellae), the flagellae without which RNA's and DNA's do not appear to operate (Grassé and Faure, 1935; Grassé, 1957, 1961).*

The image which most faithfully renders the functioning of a cell is that of a circular chain revolving on itself, with its links joining and separating at a fixed pace. The picture of a cell dominated by godlike DNA, majestic in Olympian solitude, is impressive but untrue.

In response to my interpretation, it may be objected that a virus re- produces its nucleic acid, RNA or DNA, without any activating enzyme or any other principle. Once the virion has lost its coat protein the nucleic acid replicates, finding in the host cytoplasm the nucleotides and enzymes that enable it to synthesize. For the virus, either no time factor

---

*The detection techniques employed are those of classic cytochemistry. To be sure that DNA or the RNA's do not enter into the process, we would have to use radioactive tracers, which has not been done.

interferes, or none is perceptible, which is not the case for a cellular organism with highly structured chromosomes, a genetic device obeying chronological constraints, and DNA segments deriving from their inertia either a stimulus or, less probably, the removal of an inhibition.

The dependence of DNA on genic activators, nucleotide producers, and probably also RNA producers, originally implies a dual creation of a gene plus its activator. The gene and its activator(s) contained in the nuclear sap or the cytoplasm must have been first simultaneously or consecutively obtained. Here again we see the difference between *mutation* and *innovation*. In the former the gene changes, and the corresponding activator is *always present* in the cytoplasm; the gene's expressivity is constant. In the latter it is likely possible, or at any rate, that the birth of the new gene is not synchronous with the genesis of its activators. Lacking an activator for the gene, innovation cannot occur. We sense that various possibilities have appeared in the course of time (see Chapter VIII, genetically black segments).

SYNTHESIS OF NUCLEIC ACIDS AND MOLECULAR ACQUISITION OF INFORMATION

It is worthwhile comparing the interpretations put forth here with the attempts made to achieve partial if not total synthesis of the nucleic acids and a transfer of information. We mention first the attempts to synthesize DNA *in vitro* (Kornberg, 1960, 1962, 1969, 1972). These should be capable of revealing how new information can be generated in a synthesized DNA, whose biological activity, we are bound to say, has not been demonstrated.

Kornberg *et al.* (1967) have extracted from the colon bacillus (*Escherichia coli*) an enzyme, DNA polymerase, which synthesizes chains from free nucleotides. Since this discovery, numerous experiments with DNA synthesis have been carried out.

To a buffer solution containing four kinds of deoxyribonucleotides (differing by their base: guanine, adenine, thymine, cytosine) in the forms of triphosphates (three atoms of phosphorus, instead of one), magnesium chloride, a *tiny amount of · DNA*, and a *primer RNA*, polymerase is added. This enzyme binds together the free nucleotides, the hydroxyl group of one combining with carbon C in position 3 of the deoxyribose sugar located at the end of the chain of the molecules.*

We now know that the colon bacillus has not one but three

---

*Readers wishing to know more details of the synthesis might profitably read Davidson (1972) on the biochemistry of nucleic acids.

polymerases (I, II, and III). Kornberg's polymerase (I) is now considered responsible for repairing, and not synthesizing, DNA.* Polymerases II and III are used in repairing and possibly also synthesizing DNA.

Moreover, a point highly germane to our present argument, synthesis of DNA seems to require the presence of RNA. Cavalieri (1963) established that an RNA may serve as a model in synthesis of DNA using polymerase I of *Escherichia coli*. It is now generally accepted that a priming RNA has to be present to start replication of the DNA of *E. coli* and of various phages. Kornberg's team reported highly interesting facts (Kornberg, 1972). They converted *in vitro* a single strand of DNA extracted from phage φX174 DNA to a double strand with the compulsory presence of the four kinds of deoxyribonucleoside triphosphates and one primer RNA. *Such replication implies, therefore, a covalent bond between DNA and a trigger RNA*; it is not inhibited by the antibiotic rifampicin, which does, however, block the activity of the DNA-dependent RNA polymerase. Where then does the information controlling synthesis of this RNA come from? Maybe the polynucleotide phosphorylase synthesizes it even with rifampicin present, but it is not certain.

Recently it was shown (Beljanski *et al.*, 1975) that RNA primers required for replication *in vitro* of DNA can either be synthesized under controlled conditions from ribonucleoside 5'-diphosphate by polynucleotide phosphorylase or obtained by mild degradation of *E. coli* ribosomal RNA using specific ribonucleases. Small RNA fragments (25–50 nucleotides) were characterized in several respects. They are used by DNA polymerase I for initiation of new replication sites on DNA originating from different sources.

What we note from this new addition to our knowledge is that DNA synthesis is the result of an interacting set of chemical and physical factors including the presence of free nucleotides, DNA polymerase, and a primer RNA. These conditions for DNA synthesis tell us much about its dependence on other constituents of the cell besides itself. We are a long way from that isolated position the "central dogma" claimed for it.

Cancer induced in a cell by an RNA of viral origin imposing its message on the host DNA proves that exogenous information can be overprinted on the genetic code.

---

*However, Goulian *et al.* (1967) have argued that to obtain a completely closed DNA molecule (double chain) another enzyme must be added to the polymerase: polynucleotide ligase, which binds the free ends of the molecule one to the other. The DNA used as a model came from a phage φX174 DNA. DNA synthesized *in vitro* is said to be infectious. Most biochemists treat these findings with reserve. It is hoped that their work will be confirmed by other researchers, using other materials.

VIRAL TRANSDUCTION AND TRANSFER OF ALIEN INFORMATION

The discovery of reverse transcriptase has proved that agents alien to DNA are able to supply it with fresh information. Recent observations have revealed that a virus integrated into the chromosome of a host cell may, on leaving, carry away with it a DNA segment retaining its determining ability. This phenomenon is known as viral transduction and some microbiologists (notably Anderson, 1970) attribute to it an important role in the acquisition by the living organism of extrinsic information.

Incorporation of DNA stolen from the host cell (or bacterium) in the viral DNA and its transfer to another cell (or bacterium) seem well-established facts (see Anderson, 1968). DNA segments of the host have been found in the protein envelope (or capside) of the polyoma and SV40 viruses.* Trilling and Axelrod (1970) believe that the envelopes of SV40 contain enough host DNA to synthesize four or five proteins.

Merril *et al.* (1971) introduced into a culture of fibroblasts from the skin of a man suffering from galactosemia, a characteristic of which is inability to metabolize galactose (lack of an enzyme†), λ phages carrying a *gal* operon taken from colon bacilli and producing an enzyme that turns galactose into glucose. In fibroblasts infected with the λ phages, the DNA of the *gal* operon produced an mRNA that induced production of the enzyme for eight generations of cells.

Two American microbiologists, Aposhian and Quasba, using envelopes of the virus of rodent polyoma containing host DNA labeled with a radioactive element (tritium, a hydrogen isotope), appear to have caused the DNA to be incorporated in cultured mouse embryo cells. The fact awaits confirmation; it remains to be shown that the labeled DNA is indeed found in the chromosomal DNA of the host cell.

Anderson (1970) goes so far as to explain parallel evolution by viral transduction. The eye of cephalopod mollusks is very similar in structure to that of vertebrates. One or more viruses may have transferred to the vertebrates, eye information contained in the genotype of cephalopods.

It is an ingenious hypothesis, but one of glaring fragility and improbability. Ectoparasitic insects living in mammals' fur, regardless of the order to which they belong—lice, Mallophaga, fleas, Coleoptera (*Platy-*

---

*Virus SV40 (simian virus) was discovered in cultures of renal tissue from Macaque monkeys of Indian origin.

†Specifically, α-D-galactose-1-phosphate uridyltransferase.

psyllus castoris), Forficula (*Hemimerus talpoides,* a parasite of the Gambian rat, *Cricetomys gambianus*)—have rows of strong backward-turned spines called *ctenidia*. These spines occur only in ectoparasitic insects. They turn up in orders far too phylogenetically remote from one another for them to be throwbacks. The chance of ctenidia being all derived from a single gene carried by a virus capable of being incorporated in all species is inconceivable. In any case, where do the gene(s) controlling the formation of the ctenidia come from? Similar facts can be opposed to the convergence of structure or physiology of the eyes of cephalopods and vertebrates. Anderson didn't think of the enormous number of genes controlling in one or other of these animals the formation of the eyes, their annexes, and the nerve centers corresponding to or serving them.

Known cases of viral transduction are experimental data, and we do not know whether they also occur under natural conditions. At all events the transfer of genes by the eminently pathogenic viruses appears to be possible, but only exceptionally.

One wonders whether the gene incorporated into the chromosomal DNA is lastingly tolerated by the host. Is it proof against any immunizing reaction? There is no doubt that the transmitted DNA is not a protein, but it may be treated as a foreign body by the host and consequently eliminated. Furthermore, does this DNA have any evolutionary value for the recipient? We have no facts to answer this question. The viral transduction of genes is even more aleatory than mutagenesis, which is saying a lot.

NOVELTY BY EMERGENCE

It is impossible to write about evolution without mentioning emergence. The appearance of new properties by chemical combination is the classic prototype of this phenomenon, as conceived by Morgan (1927). Kitchen salt has properties deeply different from those of its components chlorine and sodium. There is nothing mysterious about emergence; it is due to the intimate properties of the components.

It is also observed in the living plant or animal which synthesizes in practically infinite quantities the substances involved to widely differing extents in its structure, functions, and behavior. For example, synthesis of hydrochloric acid is effected by the parietal cells of the fundic glands of the stomach; the acid converts pepsinogen (a proenzyme), secreted by the chief cells of these glands, to pepsin. The enzyme owes its properties partly to the gene determining the synthesis of the propepsinogen and partly to the hydrochloric acid.

It will be remembered that DNA controls only the synthesis of proteins, the fundamental but not the sole constituents of living things. Whatever takes place in the cell after synthesis is only secondarily dependent upon the genes. Reactions between proteins, lipids, carbohydrates, mineral compounds whether ionized or not, and water in its various polymeric forms, are not under the direct influence of DNA. Heteroproteins (proteins combined in a prosthetic group) owe many of their properties to their combination with a chemical compound, the prosthetic group, as in the case of hemoglobins whose protein, a globin, is associated with a compound of ions and porphyrin. Anything that is not a protein is not encoded by DNA. In this connection remember that the cell is supplied from outside with the "materials" on which the proteins act. These materials play an important part in synthesis and metabolism. They contribute to the emergence of many properties.

Merely to consider the genes and the proteins they determine is to discover another source of emergent characteristics. Take a set of genes: Each of the proteins it produces (each its own) has its own properties. But depending on the number of grouped proteins and their mutual arrangement, the system exhibits fresh properties, just as pieces of wood, according to the way they are put together, form a framework or a fence, with different properties and functions.

The properties of a physicochemical or biological system depend upon its degree and mode of organization. As we move from the simple to the complex, new properties emerge. The human brain owes its exceptional faculties to the number and mutual arrangement of its neurons.

Biochemists have effected tremendous advances in our knowledge of cellular life, but there are still a great many important unknown factors. Numerous facts invite us, not to reject the explanations proposed, but to measure their scope and not be in too great a hurry to generalize from them. I will merely quote one case. Gramicidin (an antibiotic) is a polypeptide formed by a short chain of amino acids (10 to 40) and synthesized by spore-forming bacteria that have reached a particular stage in their life cycle. Such synthesis does not involve messenger ribonucleic acid nor ribosomes (Bhagavan et al., 1966). True, these are somewhat odd polypeptides in that they contain amino acids in the isomeric D form and amino acids such as ornithine that do not enter into the composition of the proteins. Where does the information governing synthesis of the gramicidin come from?

Some biochemists, e.g., Mach et al. (1963), maintain that the enzymes or complexes of enzymes participating in synthesis of the peptides have a structure enabling them to recognize the unfinished peptide chain and the amino acid to be incorporated in it. This may be so, but has yet to be proved. We shall wait and see. The important thing is the existence of a peptide synthesis not involving DNA. In other words, the possibilities of the cell certainly exceed those attributed to it in the present status of our knowledge.

Conclusion

By its inextricable complexity, its creations and orientations, its historicity, and, on occasion, its contradictions, evolution is quite unlike the simplified, scaled-down and totally inaccurate picture of it presented by theories. It is so vast as to make one stop and consider that its problems are very far beyond the means of present-day science. Interpretations and explanations advanced by *whomever it may be* can only be partial and tentative.

The physiologists have amply demonstrated that the living creature is an entity whose parts are all interdependent and which is forced, in order to exist, to restore at all times its physiological equilibrium by the interplay of control systems appropriate to the circumstances.

The evolutionist notes the accuracy of this classic fact, as well as that the profound and innermost evolutionary reaction is not confined to adapting the creature to its environment but commits its descendents to a certain plan of organization. This is not, we repeat, because we believe it needs to be repeated, a matter of opinion issued in a sort of spirituality which some biologists abhor, but a banal fact brilliantly confirmed by paleontology.

It is at the level of molecular phenomena, closely correlated with those unfolding at higher levels, that the mechanisms of creative evolution are situated and operate.

A cluster of facts makes it very plain that Mendelian, allelomorphic mutation plays no part in creative evolution. It is, as it were, a more or less pathological fluctuation in the genetic code. It is an accident on the "magnetic tape" on which the primary information for the species is recorded.

We must look elsewhere for the source of the evolutionary flux. The starting point of innovation is the acquisition of new genes either by addition of nucleotide sequences (codons) or by overprinting, as we imagine it to take place. In the second eventuality, creation lies in the power to recognize the enzyme activating a new sequence of codons included in the DNA molecule.

Varying ability to recognize and appraise may be regarded as a mutation of a particular, innovational type; it relates to the length and, therefore, the overall composition of the sequence of codons. In this case the emergence of new properties seems possible. The existence of enzyme mutations, as defined by us, is predictable.

In summary, the creative evolutionary process, conceived according to the data of molecular biology, involves three events:

1. Formation of a new meaningful sequence of codons
2. Formation of a specific enzyme to activate the new gene
3. Adequate identification of the enzyme depending upon cellular differentiation

In my opinion, the new information that assumes tangible form and becomes permanently integrated in the genetic code in the form of sequences of nucleotides can only be the result of preliminary intracellular operations. It is something quite different from copying errors, or from the spectrum of DNA anomalies: it is an orderly task continuing during successive generations. Such evolutionary elaboration requires a combination of precise conditions, which would seem at present to occur infrequently. Stimulation from without, internal stimulus, general reaction of the organism up to the molecular level, are likely to be the triggers of this formidable process.

But nothing of all this accounts for the orientation of evolution or the finality of the information. During ontogeny the decidely lethal genotypes are abortive; those less lethal allow development to proceed and the young or adult animal is eliminated at a late stage. I do not see what guiding power there could be for the adult form in rejection during ontogeny of genotypes causing metabolic, functional, or anatomical disturbances. That would be more a matter of regulation than of evolution. Hence the idea of a putative germinal selection must be rejected; no fact can be brought out in support of it. What reason could there be for natural selection, deemed by Weismann, Waddington, Whyte, and others insufficient to direct mutations when it affects whole populations, to become effective at the molecular level when the organism is not subject to aggression from its environment?

We are forced to admit that the determinism and mechanism of evolution involve the action of internal factors of which we have given some idea in our discussion of the acquisition of new genes.

Autoadaptive Lamarckian variation is an adequate response by the organism to aggression from the environment. How does information operate? Where does it come from? What accounts for its correspondence with the needs of the living organism? The answer in each case is silence.

Let us end our survey by drawing up a balance sheet. While still unsatisfactory, it has some favorable aspects, and dispels of one or two interpretations often presented as certainties.

1. Evolution, a guided phenomenon, is not sustained merely by random hereditary variations, sorted out by a selection operating for the good of a population.

2. Evolution demands the acquisition over time, as organisms grow more complex, of novelties whose information is inserted into the DNA strands in the form of new genes.

3. The supply of information and the subsequent creation of genes are profoundly separate mechanisms from the mutagenesis that produces alleles.

4. Paleontology reveals that lines of descent from a common stock (parent form) all show, although to unequal extents, the same propensity to achieve a given form, type, or idiomorphon.

5. Evolution in its essentials depends upon work effected at the level of infrastructures and triggered by internal and external factors, and having the effect of producing certain enzymes, probably resembling polymerases, which synthesize a new DNA and new genes by means of free nucleotides in the nuclear sap or the cytoplasm. We emphasize that the inclusion of information in the genetic code is a separate operation from its acquisition; it follows the acquisition and does not take place simultaneously with it, as does mutation. The elaboration of the information may be slow and take a great many generations; paleontology teaches us that in reality this is indeed so. Thus, DNA records and stabilizes evolution, but does not create it.

6. Mutagenesis corresponding to copying errors in the DNA is used by the organism secondarily to attain the genotype best adapted to environmental conditions. It is the main cause of differences between individuals, races, and species. If evolution takes place without the acquisition of new genes, we must assume that the first living creature contained in itself enough genes to engender, by mutation of them, all past, present, and future faunas and floras. This is absurd.

Any system that purports to account for evolution must invoke a mechanism not mutational and aleatory. This is indeed what the reformist Darwinians and Lamarckian biologists realize, hence their recourse to internal factors. The united efforts of paleontology and molecular biology, the latter stripped of its dogmas, should lead to the discovery of the exact mechanism of evolution, possibly without revealing to us the causes of the orientations of lineages, of the finalities of structures, of living functions, and of cycles. Perhaps in this area biology can go no farther: the rest is metaphysics.

Appendixes I–III

In order to make this book easier for the nonspecialist, three appendixes are included: (I) a brief classification of the animal kingdom; (II) the geological time scale; (III) a glossary of technical terms.

Appendix I
"Natural" Classification of the Animal Kingdom

UNICELLULAR ORGANISMS

1. Phylum Protozoa: This phylum consists of subphyla or classes with unknown phylogenetic relationships. It is likely to be split up into several other phyla. The fact that all the animals within it are unicellular does not imply that they are closely related. Protozoa are divided into Rhizoflagellata (Flagellata, Rhizopoda *sensu stricto*), Actinopoda, Sporozoa, Microsporidia, and Ciliata.
2. Phylum Myxosporidia: A group of parasites living as plasmodia containing ameboid germinal cells and pluricellular spores, but no spermatozoa and no ova.

DIPLOBLASTIC METAZOA

3. Phylum Porifera (sponges)
4. Phylum Cnidaria (*Hydra*, various jellyfish, corals, etc.)
5. Phylum Ctenophora (*Beroë*, Venus'girdle ...)

TRIPLOBLASTIC METAZOA

Acoelomata*

6. Phylum Platyhelminthes (flatworms, flukes, tapeworms, etc)†
7. Phylum Nemertea (*Lineus*, etc.)

*Animals not having a coelom or main body cavity.
†The phylum *Mesozoaria* is dubious. Perhaps it is a branch of Platyhelminthes.

8. Phylum Nemathelminthes (Nematodea, Gordiacea)
9. Phylum Rotifera (*Hydatina, Melicerta, . . .*)

Coelomata

10. Phylum Annelida (Class Polychaeta: nereids, sabellids, serpulids, etc.; Class Oligochaeta: earthworms, etc.; Class Hirudinea: leeches)
11. Phylum Lophophoria (Subphyla Bryozoa *sensu lato*[+] and Brachiopoda)
12. Phylum Arthropoda (Classes Trilobita,[*,+] Merostomata, Arachnida, Crustacea, Myriapoda, Insecta)
13. Phylum Echinodermata (Classes Cystoidea, [+] Blastoidea, [+] Crinoidea, Holothuroidea, Echinoidea, Asteroidea, Ophiuroidea)
14. Phylum Hemichordata (Classes Enteropneusta or Balanoglossida, Pterobranchia, Graptolita[+])
15. Phylum Pogonophora: e.g., *Siboglinum*
16. Phylum Tunicata or Urochordata (Classes Thaliacea [*Salps, Doliolum*], Ascidiacea, Appendiculata, and Larvacea)
17. Phylum Cephalochordata (Amphioxus)
18. Phylum Vertebrata (agnathans, fishes, amphibians, reptiles, birds, mammals)

A few small groups having no particular features in common with other systematic units have been given the status of phylum, but since this procedure is based on lack of knowledge it leads to uncertainty. The above list, therefore, does not include the so-called phyla Nematorhyncha, Pararthropoda, Chaetognatha, and other groups currently represented by a few genera. Only the main classes have been mentioned.

FOSSIL REPTILES (in more or less direct line of ascendance of the mammals)

Subclass Anapsida (no fenestra in the temporal region)
    Order Cotylosauria (Anapsida with the temporal region covered by 2 or 3 bones. Quadrate bone visible or concealed. Rib with one head only)
Subclass Synapsida (temporal region with a fenestra the upper edge of which is rimmed by the postorbital and squamosal)
    Order Pelycosauria (small temporal fenestra the upper edge of which is rimmed by the postorbital and squamosal and the lower edge by the jugal and squamosal. Choanae located well forward. Secondary palate. Mandible showing the beginning of an upright section. From the Carboniferous to the lower Triassic)
    Suborder Ophiacodonta
    Suborder Sphenacodonta
    Suborder Edaphosauria
    Order Therapsida (large temporal fenestra, the upper edge of which is rimmed by the postorbital and squamosal; in the most highly evolved forms with a very large

---

*According to some biologists, Bryozoa and Brachiopoda are two separate phyla. We incline to agree, but are retaining the traditional classification.
+This group is found in fossil form only.

temporal fenestra, the parietal covers only the upper edge; the lower edge or temporal arch consists of the squamosal and jugal [zygomatic arch]. Quadrate and quadratojugal very reduced. Complete secondary palate in the most highly evolved forms. Pterygoids united into a single axial bar)

Suborder Phtinosuchia: from the upper Permian in Russia

Suborder Theriodontia

Infraorder Gorgonopsia from the upper Permian (South Africa and Russia) to the Triassic

Infraorder Cynodonta from the upper Permian to the Triassic

Infraorder Tritylodonta from the upper Triassic (worldwide) to the middle Jurassic

Infraorder Therocephala from the upper Permian

Infraorder Bauriamorpha from the upper Permian to the lower Triassic (South Africa)

Infraorder Ictidosauria from the lower Liassic (South Africa)

Suborder Anomodontia

Classification of Mammals

Subclass Allotheria

Order Docodonta* : *Morganucodon* type

Order Triconodonta* : types, *Amphilestes* from the Bathonian and *Triconodon* from the Purbeckian (England)

Order Multituberculata* : types, *Plagiaulax* from the Purbeckian, *Ptilodus* from the upper and middle Paleocene (North America), *Taeniolabis* from the Paleocene (North America)

Subclass Prototheria

Order Monotremata: types, *Ornithorhynchus* (living), *Zalophus* (living), *Tachyglossus* (living)

Subclass Theria

Infraclass Pantotheria*

Order Symmetrodonta* : type *Spalacotherium* from the upper Jurassic (England)

Order Eupantotheria* : type *Amphiterium* from the Bathonian, *Dryolestes* from the Jurassic

Infraclass Metatheria

Order Marsupalia: opossums, *Marmosa,* koala, kangaroos, etc., from the upper Cretaceous to the present

Infraclass Eutheria

Superorder Carnivora

Order Creodonta*

Order Fissipedia (bears, hyenas, dogs, lions)

Order Pinnepedia (seals, walruses, sea lions)

Order Cetacea (grampuses, porpoises, sperm whales, whales, finback whales)

Superorder Protungulata

Order Condylarthra*

Order Litopterna*

Order Notoungulata*

Order Astrapotheria*

Order Artiodactyla (suiforms and ruminants)

Order Tubulidentata (aardvark)

Order Proboscidia (mastodons, elephants)
Order Hyracoidea (coney)
Order Embrithopoda*
Order Pantodonta*
Order Dinocerata*
Order Pyrotheria*
Order Xenungulata*
Order Sirenia (manatees, sea cows)
Order Perissodactyla (rhinoceros, tapirs, horses)
Order Taeniodonta*
Order Tillodonta*
Order Edentata (anteaters, armadillos, sloths)
Order Pholidota (pangolins)
Order Lagomorpha (hares and rabbits)
Order Rodentia (beavers, marmots, rats, voles, capybara, etc.)
Order Insectivora (shrews, moles, tree shrews)
Order Dermoptera (flying lemur)
Order Chiroptera (bats)
Order Primates
  Suborder Tarsioidea
  Suborder Lemuroidea
  Suborder Simioidea
  Suborder Hominoidea

*This group is found in fossil form only.

Appendix II
Geological Time Scale, Including Geological Eras and Periods

CENOZOIC	PLIOQUATERNARY		Quaternary	— 0 my
		..XLVII..........................		— 2 my
			Pliocene	— 12 my
	MIOCENE	XLVI	Pontian	
			Vindobonian	
		XLV		
			Burdigalian	— 26 my
	OLIGOCENE	XLIV		— 33 my
	EOCENE		Upper	
			Middle	—47 my
		XLIII		—52 my
			....	
			Lower	—60 my
	PALEOCENE	XLII	Thanetian	
			Montian	
			Danian	—63 my
MESOZOIC	CRETACEOUS	XLI	Senonian { Maestrichtian	—64 my
			Campanian—75 my	
			Santonian—83 my	
			Coniacian	—85 my
		XL	Turonian	—95 my
			Cenomanian	—114 my
		XXXIX	Albian	—118 my
		XXXVIII	Aptian { Urgonian	
			Barremian {	
		XXXIVII	Hauterivian } Neocomian	
			Valanginian }	—135 my

MESOZOIC	JURASSIC	MALM	XXXVI	Tithonian, Portlandian, Volgian	—136 my
			XXXV	Kimeridgian Oxfordian Upper Callovian	—139 my
		DOGGER	XXXIV	Lower Callovian (Macrocephalitian) Bathonian Bajocian	—165 my
		LIAS	XXXIII	Toarcian Pliensbachian	—170 my
	RHAETIC		XXXII	Sinemurian Hettangian	
			XXXI	Rhaetian	—180 my
	TRIASSIC		XXX	Keuper { Norian / Karnian	
			XXIX	Muschelkalk { Ladinian / Virglorian	—220 my
	PERMOTRIASSIC		XXVIII	Buntsandstein = Werfenian Tartarian = Djulfian	—240 my
UPPER PALEOZOIC	PERMIAN		XXVII	Kazanian Kungurian	—260 my
	PENNSYLVANIAN		XXVI	Autunian = Artinskian Stephanian = "Uralian"	—282 my
			XXV	Westphalian = Muskovian Upper Namurian	
	MISSISSIPIAN		XXIV	Lower Namurian { Chesterian Visean { Osagean	—320 my
			XXIII	Tournasian = Kinderhookian	—350 my
	STRUNIAN		XXII	Strunian	—360 my
	Upper		XXI	Famennian	
			XX	Frasnian	
	Middle		XIX	Givetian	
			XVIII	Couvinian (Eifelian)	—385 my
	Lower		XVII	Coblenzian { Emsian / Siegenian	—400 my
	SILURONIAN		XVI	Gedinnian = Helderbergian Ludlovian = Cayugan	—400 my

continued on next page

Upper Paleozoic, continued

LOWER PALEOZOIC				

SILURIAN
    XV   Wenlockian = upper Niagaran

(Gothlandian)
    XIV Tarannon (=Gala) = lower Niagaran —410 my
           Llandoverian (=Birkhill) = Cataract —425 my

    XIII Ashgillian   = Richmondian —440 my
           Upper Caradoc = upper Trentonian —450 my

ORDOVICIAN
    XII Lower Caradoc lower Trentonian
          Llandeilolian = Blackriverian —470 my
                   = upper Chazyan

    XI   Llandvirnian = lower Chazyan

    X   Skiddavian   = Canadian

TREMADOCIAN
    IX   Ozarkian
          Tremadocian Trempealeauan
    VIII   upper Cambrian —500 my
    VII   (= Potsdamian)

    VI   middle Cambrian —550 my
    V   (= Acadian)

CAMBRIAN
    IV

    III   lower Cambrian
    II   (= Georgian) —560 my
                         —600 my

    I   Eocambrian —680 my

PRECAMBRIAN

Notes: After H. and G. Termier. my = Millions of years ago. Paleozoic = Primary; Mesozoic = Secondary; Cenozoic = Tertiary Eras.

Appendix III
Glossary

Actinomycin An antibiotic isolated from a bacterium belonging to the order Actinomyceteales.

Allele One of the several alternative states of a functional gene unit.

Allelomorph One of two dissimilar factors which on account of their corresponding position in corresponding chromosomes are subject to alternative (Mendelian) inheritance in a diploid organism (Bateson, 1902).

Amino Acids Constituents of proteins. An amino acid may be represented as

$$R-\underset{\underset{\text{H}}{|}}{\overset{\overset{\text{NH}_2}{|}}{C}}-COOH$$

where R may be a hydrogen atom or one of 20-odd organic radicals. There are 29 amino acids, 20 of which are always present in the proteins of all living organisms. The coupling of two amino acids to form a dipeptide may be written

$$R-\underset{\underset{\text{H}_2\text{N}}{|}}{\overset{\overset{\text{H}}{|}}{C}}-\overset{\overset{\text{O}}{||}}{C}-N-\underset{\underset{\text{H}}{|}}{\overset{\overset{\text{R}'}{|}}{C}}-COOH$$

a polypeptide may be made by amino acid additions on either side, to the NH_2 group or the COOH group, or to both.

Angstrom Unit of length equal to one ten-thousandth of a micron, i.e., a billionth of a millimeter.

Apocrine gland A gland characterized by a mode of secretion halfway between that of the two kinds we describe below (holocrine and merocrine glands). The product of the secretion gathers at the apical pole of the cell and discharges into the central cavity of the gland. The axillary sweat glands belong to this type.

Apody Condition of a tetrapod vertebrate lacking limbs.

ATP Abbreviation of adenosine triphosphate, composed of one molecule of adenine (a nitrogenous purine base), one molecule of sugar (ribose), and three molecules of phosphoric acid. ATP is hydrolyzed by the enzyme adenosine diphosphatase to adenosine diphosphate (ADP) releasing a phosphoryl group and energy available for chemical reactions or for the work of muscular contraction.

Autoradiography Process in which a photographic emulsion is placed in close contact with a radioactive substance in an object (usually a thin section of organ or tissue). The radiations set free silver particles which blacken at the locations of radioactive components and thus produce an image of their shape.

Axon A long cylinderlike process, sometimes bounded by a sheath of fatty material (myelin), extending outward from the nerve cell or neuron. The complex axon and sheath form the nerve fibers.

Bacteriophage Group of viruses which parasitize bacteria. They appear in three stages: the *virion*, a free, infectious particle; the *vegetative phage* which multiplies in the host; the *prophage*, made of DNA from the RNA of the virus and fixed on the chromosome (DNA) of the bacterium. These three states have in common the same specific nucleic acid.

Blastomere A cleavage cell formed during the primary mitotic divisions of the egg.

Cheek teeth Premolars and molars which get their name from their position close to the inner surface of the cheeks.

Chert Imprecise word which generally refers to flintlike rock of chalcedony and micro-crystalline quartz.

Cistron An operational unit, equivalent to or smaller than a genetic region, controlling a specific protein.

Climax Stable culminating stage of a plant population: the cork–oak forest is the climax of the Mediterranean "maquis" or "garrigues" community; the so-called primary equatorial forest is the climax of forested populations of equatorial areas; the beech-sugar maple forest is the climax of the forest in Indiana; the *Stipa–Sporobolus–Bouteloua* grassland is a climax community in Iowa. The word climax also applies to animal populations under stable environmental conditions.

Codon A sequence of three contiguous nucleotides determining the code of an amino acid or of a chain termination.

Convergence Morphological or physiological similarity in distantly related forms.

Creatine Synonym for methylguanidineacetic acid; found in the muscles of vertebrates as creatinine. It is a product of protein degradation, excreted in urine.

Cryptic species Species morphologically indistinguishable but incapable of interbreeding. They differ either by their habits, or, more often, by the number or structure of their chromosomes.

Diploblastic Multicellular organisms having only two germ layers (ectoderm and endoderm); they comprise the sponges, Cnidaria, and Ctenophora.

Diploid Having a double set of chromosomes ($2n$); the normal chromosome number of the cells of a particular organism derived from a fertilized egg.

DNA Abbreviation of deoxyribonucleic acid, the material which transmits hereditary information (genetic code); a major component of chromosomes. It is a polymer of nucleotides, which result from the esterification by a molecule of phosphoric acid of a heteroside called nucleoside which is a compound formed by a heterocyclic nitrogenous base (purine or pyrimidine) and a sugar at C-5. The composition of the combined nucleotides varies depending on their bases.

DNA polymerase The enzyme which catalyzes the formation of DNA from deoxyribonucleoside 5′-triphosphates, using DNA as a template.

Dollo's rule The principle that evolution is irreversible to the extent that structures or functions once lost cannot be regained.

Ecomorphosis Physical or functional modification related to environmental changes.

Endemism Confinement of certain animal or plant species to a well-delimited geographical area. **Neoendemism** refers to those species which have developed within the area they now live in, after they became isolated (such as the Galapagos finches). **Paleoen-**

demism refers to relict species, which used to thrive and which now subsist in some refuges (such as the tuatara, specific to New Zealand islets).

Enzyme (or **diastase**) A protein molecule capable of catalyzing a certain cellular reaction and which is left unchanged once the reaction is finished.

Endomitosis Multiplication of chromosomes in a nucleus without subsequent nuclear division. The polyploidization of the cell (multiplication by n or $2n$ chromosomes) is usually induced by endomitosis.

Episome A particle of genetic significance which is observed in bacteria outside the normal chromosome or within it, and which did not derive from a mutation of bacteria, but from external causes. The F^+ factor, which plays a role in the sexuality of the colon bacillus, and the DNA of the lysogene bacteriophages are both eipsomes.

Ergastoplasm Cell organelle composed of vesicles that are usually flat and irregular in shape; it comes into play in numerous syntheses. Often associated with ribosomes.

Fetalization Theory mainly conceived by the Dutch anatomist, Bolk, and which maintains that evolution is achieved by the maintenance of fetal characters in the adult. To carry this idea to the extreme, one could say that man is nothing but the fetus of an ape that has acquired the capacity to reproduce itself.

Fibroblast A fusiform cell which is part of the connective tissue of the embryo. In tissue cultures, certain cells change, partly lose their properties, and become spindle-shaped (for example, cultures of fibroblasts taken from the heart of a chicken embryo).

β-Galactidase An enzyme that hydrolyzes lactose into glucose and galactose.

Galactoside A glycoside that yields galactose which is a stereochemical isomer of glucose and does not exist in the free state in living beings.

Genetic code Complex of codons contained in chromosomal as well as mitochondrial DNA, which determines the morphological, physiological, and ethological character of living beings.

Genetic drift Genetic changes in populations caused by random phenomena (loss of genes) rather than by selection.

Geniculate bodies Nervous centers located in the lower brain region in the optic layers. The two external geniculate bodies are connected with the arms of the anterior corpora quadrigemina; the two internal geniculate bodies are connected with the posterior corpora quadrigemina. The external geniculate body (one on each side) receives 80% of the visual fibers; the remaining 20% run into the *pulvinar,* a center located in the thalamus, deriving from the embryonic diencephalon.

Genome The sum total of genetic material in the chromosomes of a zygote.

Genotype The totality of the genetic material of cells; the total genetic endowment of an individual at a given locus (Johannsen, 1911).

Globin A protein containing histidine and lysine, often combined with an iron compound (such as protoheme in the case of hemoglobin).

Golgi apparatus A cell organelle consisting of individualized units called dictyosomes which are made up of a series of flattened "saccules"; they produce vesicles which bud off into the surrounding cytoplasm. The Golgi apparatus plays a part in the various chemical mechanisms of the cell; the Golgi apparatus is absent in schizophytes and in a few protozoans.

Gradient Spatial distribution of a substance or of a property in an ovum, in an organism, or in an organ, starting from a maximum value (head of the gradient) and regularly decreasing to a minimum value. The gradient is distributed along the main axis of the being or organ considered. The distribution area of the substance or of the property constitutes a gradient field. In the ovum, the gradients are distributed along the main axis going from the animal pole to the vegetal pole.

Heterozygote   A zygote derived from the union of gametes dissimilar with respect to the constitution of their chromosomes (Bateson, 1902).

Holocrine gland   A gland in which the product of the secretion fills the cells which then degenerate and set the substance free (e.g., sebaceous glands of the mammalian skin).

Homologous chromosomes   Chromosomes which pair (the genes of a pair lining up opposite each other) and which bear identical genes at the same genetic loci.

Homozygote   A zygote derived from the union of gametes with identical genes (*Bateson*, 1902).

Hypertely   (*Hyper*: above; *telos*: goal, end). Extreme state of an organ or function which places the organism or its lineage in unfavorable conditions.

Imaginal   Of or relating to the adult insect (imago).

Incus   Small bone of the middle ear of mammals, also called the anvil.

Krebs cycle, also called **citric acid cycle**   This cycle occurs during cellular respiration in the mitochondria of cellular organisms. It metabolizes glucose, and it consists of two main phases: (1) transformation of glucose into pyruvic acid, and (2) oxidation of this acid through a long series of reactions, producing 5 molecules of water and 3 molecules of carbonic anhydride.

Lethal factors   Genetic factors which render an organism nonviable.

Lethality   The quality or state of being lethal. A lethal gene places its carrier in a pathological state. In double dose (a maternal lethal gene and a paternal lethal gene for a given pair of genes or allelomorph pair), it causes death.

Ligase   An enzyme catalyzing the formation of a phosphodiester bond between the free end 5'-phosphate of an oligo- or polynucleotide and the group 3'-OH of another oligo- or polynucleotide located in its close vicinity.

Luteinizing cell   Cell of the corpus luteum of the ovary containing yellowish fat and the so-called luteinizing hormone or *progesterone,* which in mammals determines the modifications of the uterus, preparing it and making gestation possible.

Lysosomes   Vesicular corpuscles scattered in the cytoplasm, surrounded by a lipoprotein membrane and storing enzymes which usually have lytic properties. The digestion of the membrane liberates the enzyme.

Malleus   Ossicle of the middle ear of mammals articulating with the head of the incus; also called the hammer.

Meiosis   A form of mitosis in which the nucleus divides twice and the chromosomes once. The prophase of meiosis is the prophase of the first of the two divisions (Farmer and Moore, 1905).

Meiotic reduction   Same as meiosis.

Merocrine gland   A gland in which the product of the secretion accumulates in the cytoplasm of the cells forming the gland; it is discharged with or without the breaking down of the apical pole, and the secretory activity can recur.

Messenger RNA (mRNA)   RNA serving as a template for protein synthesis and as a link between DNA and ribosomes. There are two other major types of RNA: ribosomal RNA (rRNA) and transfer RNA (tRNA).

Metabolite   A substance essential to the metabolism of a particular organism or metabolic process.

Mitochondria   Semiautonomous cell organelles scattered in the cytoplasm, containing enzymes, adenosine triphosphate (ATP), and DNA. They perform various functions, notably the delivery of energy by means of their ATP. They measure about 1 μm, but can reach greater lengths; they may be shuttle- or rod-shaped (Benda, 1897).

Mitosis   The process by which daughter chromosomes are separated into two identical groups; the diagnostic property of division of the nucleus. The term may be conveniently used in contradistinction to meiosis. Each daughter cell receives $2n$ chromosomes (Flemming, 1882).

Monodactyly   The state of a limb having only one digit.

Monophyly   The derivation of one or more lineages from the same ancestral organisms. In opposition to di- or polyphyly.

Mutation   Any physical or functional, heritable variation. Two types of mutations are distinguished: *gene mutations* due to changes in the genetic code, and *chromosomal mutations* resulting from modification of the order of the genes (e.g., inversion of a segment) in a chromosome or from a numerical variation involving some part of the characteristic chromosomes of the species.

Mutation pressure   The continued recurrent production of an allele by mutation tending to increase its frequency in the gene pool of a population.

Oncornavirus   A carcinogenic virus, the genetic material of which is composed of RNA.

Ontogenesis or **Ontogeny**   The course of development of an individual organism from the ovum to the adult form. Some authors incorrectly use this word as a synonym of embryogeny (i.e., the formation or development of the embryo).

Operator   A gene or group of genes capable of interacting directly or indirectly with a specific repressor, thereby controlling the type of regulation of the adjacent operon.

Organelle or **Organite**   A structure of characteristic morphology and function within the cell; the analogue of an organ in a multicellular organism; e.g., Golgi apparatus, mitochondrion.

Orthogenesis   Evolution with successive stages, established following geological chronology, which show a regular progression or regression in some predestined direction. The evolution of horses' hooves and of deer's antlers during the Tertiary are two types of orthogenesis (as defined by Cuénot, 1946).

Panchronic species (or **groups**)   Old or very old species or groups which reproduce without major change.

Panmixia   Participation in sexual reproduction of all the individuals of one population entirely at random and in the absence of any selection (resulting ultimately in a high degree of uniformity if the population is strictly closed).

Phage   A common abbreviation of **bacteriophage**.

Phenotype   The external appearance produced by the reaction of an organism of a given genotype with a given environment (Johannsen, 1911).

Phosphagens   Name given by Eggleton (1929) to phosphate compounds of creatine [phosphocreatine or phosphorylcreatine, adenosine triphosphate (ATP)] occurring in muscles and other organs, where they play a very important role in hydrolysis by releasing a molecule of phosphoric acid and energy. They play a role in the chemistry of muscular contraction.

Phosphocreatine   See phosphagens and creatine.

Phylogeny   The origin and evolution of a species, genus, family, order, class, or phylum (commonly depicted by a family tree or genealogical tree).

Pleiotropy   Capacity of one particular gene to produce more than one effect; i.e., having multiple phenotypic expressions.

Polydactyly   The condition of having more than the normal number of toes or fingers.

Polyphyletic   A term applied to a systematic unit derived from two or more sources.

Polyploid   An organism with more than two sets of homologous chromosomes. The terms used are triploid, tetraploid, pentaploid, hexaploid, heptaploid, octoploid (for octaploid), nonaploid (for enneaploid), decaploid, undecaploid (for hendecaploid), dodecaploid, and so on. Higher mutiples are best referred to as 14 x, 22 x, and so on (Winkler, 1916).

Polysomy   Chromosomal anomaly with one or more chromosomes represented three or more times instead of twice. Mongoloid idiocy corresponds to a trisomy of chromosome 21.

Position effect   The differences in effect of two or more genes according to their distances apart in the chromosome strand (Sturtevant, 1926).

Proteins   Nitrogenous organic compounds made up of amino acids linked together in the form of long chains, which can be combined with various chemical groups. They are the most important constituents of living beings.

Pseudopods   Protrusions (with many possible shapes) of amebas and amebocytes and of numerous free cells, serving as organs of locomotion or for taking up living prey or inert food.

Regulating mechanism   The mechanism which restores and maintains the normal state of the living being when it is modified; the regulation can be of a morphological or chemical nature. It affects all the activities of the living being, including behavior.

Regulatory gene   A gene which controls the synthesis of the products of one or several other genes.

Ribosomes   Cellular particles scattered in cytoplasm and made up of RNA (ribosomal RNA) and proteins; the primary site of protein synthesis.

Rifampicin   An antibiotic isolated from a *Streptomyces* bacterium belonging to the order Actinomycetales.

RNA   Abbreviation of ribonucleic acid; this nucleic acid differs from DNA in that it contains another sugar—D-ribose or ribofuranose—and some other bases.

RNA polymerase   The enzyme which catalyzes the formation of RNA from ribonucleoside 5'-triphosphates, using DNA as a template.

Selection coefficient   A quantitative measure of the intensity of selection expressed as s, the selection coefficient, which is the proportional reduction in the gametic contribution of a particular genotype compared with a standard genotype, usually the most favored. The contribution of the favored genotype is taken to be 1, and the contribution of the genotype selected against is then $(1 - s)$.

Selection pressure   Statistical force leading to the change of allelic frequencies due to the influence of natural selection, i.e., possible differential reproduction of the various genotypes (Professor P. L'Héritier). The intensity of natural selection is usually measured by the change of gene frequency per generation due to the influence of selection.

Somation   Physical or functional variation of the body or *soma* which does not affect the germ cells or *germen*. Acquired characteristics are connected with environmental effects. They are not hereditary.

Speciation   The ultimate splitting of a phyletic line; the process of formation of species.

Stapes   Ossicle of the middle ear of mammals; also called the stirrup.

Striated zone of the occipital lobe   The occipital lobe is located at the posterior end of the cerebral hemisphere; it mainly performs a visual function. The striated zone or area lies in between the two sides, at the bottom of the calcarine fissure (main horizontal fissure in the mesial surface of the occipital lobe) and on the external side of the occipital pole. Each point of the retina is projected on a corresponding point of the cerebral cortex which forms locally a "cortical retina."

Suppressor gene   A gene which reverses or cancels the action of another gene or genes.

Teratology   A branch of zoology concerned with the study of monsters and their genesis.

Thymidine   One of the nucleosides making up DNA and containing a pyrimidine base, thymine.

Tribosphenic molar   A mammalian molar, originally bearing three tubercles, and later in evolution five, and which in occlusion, by fitting with or opposing its mate, can grind as well as cut food.

Triploblastic   Animals having three germ layers, the ectoderm, endoderm, and mesoderm.

Tritium or ^{3}H   The radioactive isotope of hydrogen that has atoms of three times the mass of ordinary light hydrogen atoms, that has a half-life of about 12.5 years and emits radiation. It is used widely as a "label" to follow the changes occurring in substances within an organism.

Viruses   Infective agents, capable of multiplication only in living cells. Their size is much smaller than that of a bacterium; their genetic material is composed of a single nucleic acid (DNA or RNA) and one protein. Examples are the tobacco mosaic and poliomyelitis viruses.

Viviparity   Mode or reproduction in which the young develops within the body of its mother and is produced completely formed (a few arthropods, a few amphibians and reptiles, and all mammals except for monotremes).

X Chromosome, Y chromosome, or **heterochromosomes**   Chromosomes determining sex. In man and *Drosophila,* XX = female, XY = male.

Zygote   The cell formed by the union of two gametes of opposite sex; the fertilized ovum (cf. heterozygote, homozygote).

Bibliography

An Introduction to the Study of Evolution

Alvarado, R., *et al.* (1959). La teoria de la evolucion a los cien años de la obra de Darwin. *Rev. Univ. Madrid* **8,** 9–559.

Anthony, R., and Cuénot, L. (1939). Enquête sur le problème de l'hérédité conservatrice; les callosités carpiennes du Phacochère. *Rev. Gen. Sci. Pures Appl.* **50.** 313–320.

Baldwin, J. M. (1902). "Development and Evolution." New York and London.

Berg, E. S. (1926). "Nemogenesis or Evolution Determined by Law." Constable, London.

Bergson, H. (1907). "L'évolution créatrice." Alcan, Paris.

Bolk, L. (1926). "Das Problem der Menschwerdung." Jena.

Caullery, M. (1931). "L'Évolution." Payot, Paris.

Colbert, E. A., (1955). Evolution of the horned dinosaurs. *Evolution* **5,** 145–163.

Cope, E. D. (1896). "The Primary Factors of Organic Evolution." Chicago.

Darwin, C. (1859). "The Origin of Species." London [John Murray (1964 Facsimile), Harvard University Press, Cambridge, Mass.].

Darwin, C. (1871). "The Descent of Man." London.

Darwin, C., and Wallace, A. R. (1858). On the tendency of species to form varieties; and on the perpetuation of varieties and species by natural means of selection. [Note communication by Sir Charles Lyell and Sir Joseph Hooker (lecture given July 1, 1858)]. *In* "Evolution by Natural Selection," (Gavin de Beer, ed.), 257–263, Cambridge University Press, Cambridge, Mass.

Delage, Y. (1894). La structure du cytoplasme et l'hérédité. *Anneé Biol.* **2,** 495.

Delage, Y. (1903). "L'hérédité et les grands problèmes de la biologie générale," 2nd Ed. Schleicher, Paris.

Delage, Y., and Goldsmith, M. (1909). "Les théories de l'évolution." Flammarion, Paris.

Dobzhansky, T. (1941). "Genetics and Origin of Species." Columbia University Press, New York.

Dobzhansky, T., and Boesiger, E. (1968). "Essais sur l'évolution," Les grands problèmes de la biologie", Monogr. No. 9. Masson, Paris.

Dollo, L. (1893). Les lois de l'évolution. *Bull. Belge Geol.* **7,** 164–167.

Driesch, H. (1909). "Philosophie der Organismen." "La philosophie de l'organisme" (Fr. transl. by Kollmann). Rivière, Paris, 1921.

Eimer, T. (1897). "Orthogenesis der Schmetterlinge" (with the collaboration of C. Fickert). Engelmann, Leipzig.

262

Fox, S. W. (1965). "The Origin of Prebiological Systems and of Their Molecular Matrices." Academic Press, New York.

Gaudry, A. (1896). "Essai de paléontologie philosophique." Baillière, Paris.

Geoffroy Saint-Hilaire, E. (1818). "Philosophie anatomique." (2 vol.) Paris.

Gilson, E. (1971). "D'Aristote à Darwin et retour." Vrin, Paris.

Goldschmidt, R. (1944). "The Material Basis of Evolution," 4th Ed. Yale Univ. Press, New Haven, Connecticut.

Goldschmidt, R. (1952). Evolution as viewed by one geneticist. *Am. Sci.* **40**, 84–135.

Grassé, P.-P. (1943). "L'Évolution: faits, expériences, théories." Cent. Doc. Univ., Paris.

Grassé, P.-P. (1959). Les incertitudes des doctrines évolutionnistes. *Rev. Univ. Madrid* Nos. 29–31.

Grassé, P.-P. (1966). Évolution. *In* Précis de Biologie Générale, pp. 753–963. Masson, Paris.

Greenwood, T. (1948). Le principe de l'évolution émergente dans la philosophie anglaise contemporaine. *Etud. Philos. N. Ser.* **3**. 70–92.

Gurwitsch, A. (1914). On practical vitalism. *Am. Nat.* **49**. 763–770.

Guye, C. E. (1942). "L'évolution physico-chimique," 2nd Ed. Hermann, Paris.

Guyénot, E. (1939). La véritable pensée de Lamarck et le transformisme contemporain. *Scientia*, Bologna **66**. 175–183.

Haeckel, E. (1866). "Generelle Morphologie der Organismen," Vols. I and II. G. Reimer, Berlin.

Haeckel, E. (1874). "Histoire de la création naturelle." Reinwald, Paris.

Haldane, J. B. S. (1924). A mathematical theory of natural and artificial selection. *Trans. Cambridge Philos. Soc.* **23**, 19–41.

Haldane, J. B. S. (1932). "The Causes of Evolution." Longmans and Green, New York and London.

Haldane, J. B. S. (1955). The measurement of natural selection. *Atti, Congr. Int. Gene. Caryologia, 9th, Florence* Suppl., 480–487.

Haldane, J. B. S. (1958). The theory of evolution, before and after Bateson. *J. Genet.* **56**, 11–27.

Hawkes, J. G., ed. (1968). "Chemotaxonomy and Serotaxonomy," The Systematics Association, Spec. Vol., No. 2. Academic Press, New York.

Hovasse, R. (1943). "De l'adaptation à l'évolution par la sélection." Hermann, Paris.

Hovasse, R. (1950). "Adaptation et évolution." Hermann, Paris.

Huxley, J. (1942). "Evolution, the modern synthesis." G. Allen and Unwin, (First edit.,) London. (2nd Ed.: 1944.)

Huxley, J. (1953). "Evolution in Action." Chatts and Winders, London.

Jeannel, R. (1942). "La genèse des faunes terrestres." Presse Univ. de France, Paris.

Jeannel, R. (1950). "La marche de l'évolution," 171 pp. Muséum, Paris.

Jeschikov, J. J. (1937). Zur Rekapitulationslehre. *Biol. Gen.* **13**, 67–100.

King, J. L., and Jukes, T. (1969). Nondarwinian evolution. *Science* **164**, 788–797.

Lamarck, J.-B. Monnet de (1809). "Philosophie zoologique." (1873) Ed. Savy, Paris.

Lebedkin, S. (1937). The recapitulation problem. I and II. *Biol. Gen.* **13**, 391–417, 516–594.

Lehman, J. P. (1973). "Les preuves paléontologiques de l'évolution." Presses Univ. de France, Paris.

Le Roy, E. (1927). "L'exigence idéaliste et le fait de l'évolution." Boivin, Paris.

Lévine, R. P. (1969). "Génétique." Ediscience, Paris.

Mayr, E. (1963). "Animal Species and Evolution." Harvard Univ. Press, Cambridge, Massachusetts.

Mayr, E. (1969). "Principles of Systematic Zoology," 428 pp. McGraw-Hill, New York.

264 Bibliography

Mazenot, G. (1940). La "loi" d'accélération phylogénétique ou de la précession des caractères (loi de A. Pavlov). *Bull. Soc. Linn. Lyon* Nos. 1–3, p. 19–56.
Meyer, F. (1954). "La problématique de l'évolution." Presses Univ. de France, Paris.
Mora, P. T. (1965). The Folly of Probability. *In* "The Origin of Prebiological Systems." (W. Fox, ed.), Academic Press, New York.
Morgan, C. L. (1927). "Emergent Evolution," The Gifford Lectures, 1922. Holt, New York.
Morgan, T. H. (1932). "The Scientific Basis of Evolution." Norton, New York.
Neuville, H. (1927). De certains caractères de la forme humaine et de leurs causes. *Anthropologie* **37**, 305–328.
Osborn, H. F. (1921). "L'origine et l'évolution de la vie" (Fr. transl. by F. Sartiaux). Masson, Paris.
Rensch, B. (1959). "Evolution Above the Species Level." Methuen, London.
Rosa, D. (1931). "L'ologenèse. Une nouvelle théorie de l'évolution." Payot, Paris.
Schindewolf, O. H. (1950). "Grundfragen der Paläontologie." Schweizerbart, Stuttgart.
Sewertzoff, A. N. (1931). "Morphologische Gesetzmässigkeiten der Evolution." Fischer, Jena.
Simpson, G. G. (1944). "Tempo and Mode in Evolution." Columbia Univ. Press, New York.
Strickberger, M. W. (1968). "Genetics," 868 pp. Macmillan, New York.
Teilhard de Chardin, P. (1957). Les fondements et le fond de l'idée d'évolution. *In* "La Vision du Passé." 163–197, Le Seuil, édit. Paris.
Vandel, A. (1958). "L'Homme et l'évolution," 2nd Ed. Gallimard, Paris.
Vandel, A. (1968). "La Genèse du vivant," *Les Grands Problèmes de la Biologie,* Monogr. No. 6, Masson, Paris.
Vernet, G. (1950). "L'Évolution du Monde Vivant." Plon, Paris.
Vialleton, L. (1908). "Un problème de l'évolution." Coulet, Montpellier.
von Bertalanffy, L. (1948). "Les problèmes de la vie" (Fr. transl. by M. Deutsch). Gallimard, Paris, 1961.
von Bertalanffy, L. (1951). "Theoretische Biologie," 2nd Ed., 2 Vols., Bern.
Wallace, A. R. (1891). "Le darwinisme, exposé de la théorie de la sélection naturelle avec quelques-unes de ses applications" (Fr. transl. by H. de Varigny), 674 pp. Paris.
Wintrebert, P. (1962). "Le Vivant Créateur de son Evolution." Masson, Paris.
Wright, S. (1969). "Evolution and the Genetics of Populations," 3 Vols. Univ. of Chicago Press, Chicago, Illinois.

Large Treatises and Reference Books

Boureau, E. (1964). "Traité de paléobotanique," 4 vols. Masson, Paris.
Chadefaud, M., and Emberger, L. (1960). "Traité de botanique systématique," 3 vols. Masson, Paris.
Grassé P.-P., ed. (1948–1972). "Traité de Zoologie," 35 vols. Masson, Paris.
Moore, R. C. (1953). "Treatise on Invertebrate Paleontology." Geol. Soc. Am., Univ. of Kansas Press, Lawrence.
Piveteau, J., ed. (1952–1966). "Traité de Paléontologie," Vols. I–VII (10 vols.). Masson, Paris.

Chapter I From the Simple to the Complex—Progressive Evolution, Regressive Evolution.

Barghoorn, E. S. (1971). The oldest fossils. *Sci. Am.* **224** (5), 30–42.
Barghoorn, E. S., and Schopf, J. W. (1966). Microorganisms three billion years old from the precambrian of South Africa. *Science* **152**, 758–763.

Dauvillier, A. (1958). "L'origine photochimique de la vie." Masson, Paris.

Grassé P.-P. (1952). Embranchement des Protozoaires. Rapports systématiques et affinités. In "Traité de Zoologie" (P.-P. Grassé, ed.), Vol. I, Part 1, pp. 39–52. Masson, Paris.

Grassé, P.-P. (1969). Les Protozoaires sont-ils les ancêtres des Métazoaires? Symp. Int. Zoofil. 1st Salamanca Univ. pp. 64–92.

Grassé P.-P. (1971). "Toi, ce Petit Dieu," Albin Michel, Paris.

Guibé J. (1970). La réduction des membres chez les Reptiles. In "Traité de Zoologie" (P.-P. Grassé, ed.), Vol. XIV, Part 2, pp. 194–201. Masson, Paris.

Haeckel, E. (1882). "Les preuves du transformisme, réponse à Virchow" (J. Soury, transl.) 2nd Ed. Baillière, Paris. (Orig. Germ. Ed.: 1877.)

Jefferies, R. P. S. (1968). The subphylum Calcichordata (Jefferies, 1967), primitive fossil Chordates with Echinoderm affinities. Bull. Br. Mus. (Nat. Hist.), Geol. 16, 243–339.

Jefferies, R. P. S., and Prokop, R. J. (1972). A new Calcichordate from the Ordovician of Bohemia and its anatomy, adaptations and relationships. Biol. J. Linn. Soc. 4, 69–115.

Raynaud, A. (1963). La formation et la régression des ébauches des membres de l'embryon d'Orvet (Anguis fragilis). Bull. Soc. Zool. Fr. 88, 299–324.

Raynaud, A. (1972). Morphogenèse des membres rudimentaires chez les Reptiles: un problème d'embryologie et d'évolution. Bull. Soc. Zool. Fr. 97, 469–485.

Saint-Seine, P. (1951). Les fossiles au rendez-vous du calcul. Congr. Int. Philos. Sci. pp. 75–84.

Schopf, J. W. (1970). Precambrian micro-organisms and evolutionary events prior to the origin of vascular plants. Biol. Rev. Cambridge Philos. Soc. 45, 319–352.

Scourfield, D. J. (1937). An Anomalous Fossil Organism, Possibly a New Type of Chordate from the Upper Silurian of Lesmahagow, Lanarkshire, Ainiktozoan loganense gen. et sp. nov. Proc. R. Soc. London (B) 121, 533–547.

Ubaghs, G. (1971). Diversité et spécialisation des plus anciens Échinodermes que l'on connaisse. Biol. Rev. Cambridge Philos. Soc. 46, 157–200.

Von Wettstein, D. (1957). Genetics and the sub-microscopic cytology of plastids. Hereditas 43, 303–317.

White, E. I. (1946). Yomoytius kerwoodi, a new Chordate from the Silurian of Lanarkshire. Geol. Mag. 83, 89–97.

Chapter II Creative Evolution or the Appearance of Types of Organization

Barghusen, H. R. (1968). The lower jaw of Cynodonts (Reptilia, Therapsida) and the evolutionary origin of mammal-like adductor jaw musculature. Pastilla (Peabody Mus. Nat. Hist. Yale Univ.) 116, 1–49.

Crompton, A. W. (1963). On the lower jaw of Diarthrognathus and the origin of the mammalian lower jaw. Proc. Zool. Soc. London 140, 697–753.

Crompton, A. W., and Jenkins, F. A. (1968). Molar occlusion in late Triassic mammals. Biol. Rev. Cambridge Philos. Soc. 43, 427–458.

de Ricqlès, A. (1972a). Vers une histoire de la physiologie thermique. Les données histologiques et leur interprétation fonctionnelle. C. R. Acad. Sci., Sér. D, 275, 1745–1748.

de Ricqlès, A. (1972b). Vers une histoire de la physiologie thermique. L'apparition de l'endothermie et le concept de Reptile. C.R. Acad. Sci., Sér. D, 275, 1875–1878.

Devillers, C. (1961a). Ictidosauria. In "Traité de Paléontologie" (d. Piveteau, director), Vol. VI, Part 1, pp. 192–223. Masson, Paris.

Devillers, C. (1961b). Origine de l'oreille moyenne des Mammifères. In "Traité de Paléontologie" (J. Piveteau, ed., Vol. VI, Part 1, pp. 371–407. Masson, Paris.

Edinger, T. (1948). Evolution of the horse brain. Mem. Geol. Soc. Am. 25, 1–177.

Ginsburg, L. (1970). Les Reptiles fossiles. *In* "Traité de Zoologie" (P.-P. Grassé, ed.), Vol. XIV, Part 3, pp. 1188–1332. Masson, Paris.

Goldschmidt, R. (1944). Cryptic bobbed alleles in *Drosophila melanogaster*. *Am. Nat.* **78,** 564–568.

Haeckel, E. (1866). "Generelle Morphologie der Organismen," 2 vols. Reimer, Berlin.

Hopson, J. A. (1971). Postcanine replacement in the gomphodont cynodont *Diademodon*. Early Mammals. *Zool. J. Linn. Soc.* **50,** Suppl. 1, 1–21.

Hopson, J. A., and Crompton, A. W. (1969). The origin of mammals. *Evol. Biol.* **3,** 15–72.

Howell, A. B. (1937). Morphogenesis of shoulder architecture. Part V. Monotremata. *Q. Rev. Biol.* **12,** 191–205.

Jarwik, E. (1960). "Théories de l'évolution des Vertébrés." Masson, Paris.

Kermack, K. A. (1967). The interrelations of early mammals. *Zool. J. Linn. Soc. London,* **47,** 241–249.

Kermack, D. M., and Kermack, K. A., eds. (1971). Early Mammals. *Zool. J. Linn. Soc. London* **50,** Suppl. 1.

Kermack, K. A., and Kielan-Jaworowska, Z. (1971). Therian and nontherian mammals. Early Mammals. *Zool. J. Linn. Soc. London* **50,** Suppl. 1, 103–115.

Krebs, B. (1971). Evolution of the mandible and lower dentition in dryolestids. (Pantotheria, Mammalia). Early Mammals. *Zool. J. Linn. Soc. London* **50,** Suppl. 1, 89–102.

Kühne, W. G. (1956). "The Liassic Therapsid *Oligokyphus*." British Museum, London.

Lehman, J.-P. (1959). "L'évolution des Vertébrés inférieurs," Quelques problèmes, Monograph. Dunod, Paris.

Massoud, Z. (1967). Contribution à l'étude de *Rhyniella praecursor,* Collembole fossile du Dévonien. *Rev. Ecol. Biol. Sol* **4,** 497–505.

Mayr, E. (1963). "Animal Species and Evolution," 797 pp. Belknap Press of Harvard University, Cambridge, Massachusetts.

Mendrez, C. (1972). Premières ébauches d'un palais secondaire osseux chez les Reptiles mammaliens. *C.R. Acad. Sci. Sér. D,* **274,** 2960–2961, Paris.

Mills, J. R. E., (1971). The dentition of *Morganucodon*. Early Mammals. *Zool. J. Linn. Soc. London* **50,** Suppl. 1, 29–63.

Olson, E. C. (1944). Origin of mammal based on the cranial of therapsid suborders. *Spec. Pap. Geol. Soc. Am.* **55,** 1–136.

Osborn, H. F. (1936). "Proboscidea. A Monograph of the Discovery, Evolution, Migration and Extinction of the Mastodonts and the Elephants of the World," Vols. I and II. *Am. Mus. Nat. Hist.,* New York.

Romer, A. S. (1970). The Chañares (Argentina) triassic reptile fauna. VI. A Chiniquidontid Cynodon with an incipient squamosal-dentary jaw articulation. *Breviora* No. 344, 18 pp.

Simpson, G. G. (1928). "A Catalogue of the Mesozoic Mammalia in the Geological Department of the British Museum." British Museum, London.

Simpson, G. G. (1951). "Horses." Oxford Univ. Press, London and New York.

Simpson, G. G. (1959). Mesozoic mammals and the polyphyletic origin of mammals. *Evolution* **13,** 405–414.

Simpson, G. G. (1960). Concluding remarks: Mesozoic mammals revisited. Early Mammals. *Zool. J. Linn. Soc. London* **50,** Suppl. 1, 181–198.

Störmer, L. (1955). "Merostomata," Treatise on Invertebrate Paleontology, pp. 4–41. *Geol. Soc. Am. Univ. of Kansas Press,* Lawrence.

Termier, H., and Termier, G. (1968). "Biologie et écologie des premiers fossiles," Les Grands Problèmes de la Biologie (P.-P. Grassé, ed.). Masson, Paris.

Vandebroek, G. (1969). "Evolution des Vertébrés de leur origine à l'Homme," 583 pp. **94,** 117–160.

Vandebroek, G. (1969). "Evolution des Vertébrés de leur origine à l'Homme," 583 pp. Masson, Paris.

Watson, D. M. S. (1931). On the skeleton of a Bauriamorph reptile. *Proc. Zool. Soc. London* **2**, 1163–1205.

Watson, D. M. S. (1954). On *Bolosaurus* and the origin and classification of reptiles. *Bull. Mus. Comp. Zool. Harvard Univ.* **111**, 297–449.

Chapter III Evolution—A Discontinuous Historical Phenomenon

Bailey, D. W. (1956). Re-examination of the diversity in *Partula taeniata*. *Evolution* **10**, 360–366.

Bauchot, R. (1972). Le degré d'organisation cérébrale des Mammifères. *In* "Traité de Zoologie" (P.-P. Grassé, ed.), Vol. XVI, Part 4, pp. 360–384. Masson, Paris.

Chaline, J. (1972). Les Rongeurs du Pléistocène Moyen et Superieur de France, "Cahiers de Paléontologie." CNRS, Paris.

Chaline, J. (1973). Les Rongeurs fossiles au service de la préhistoire. *Recherche* **4**, 180–181.

Crampton, H. E. (1916). Studies on the variation, distribution, and evolution of the genus *Partula*. The species inhabiting Tahiti. *Carnegie Inst. Washington Publ.* **228**, 1–311.

Crampton, H. E. (1925). Contemporaneous organic differenciation on the species of Partula living in Moorea, Society Islands. *Am. Nat.* **59**, 5–35.

Crampton, H. E. (1932). Studies on the Variation, Distribution, and Evolution of the Genus *Partula*. *Publ. Carnegie Inst.*, No. 410, 1–335.

Cuénot, L. (1952). Phylogenèse du Règne animal. *In* "Traité de Zoologie" (P.-P. Grassé, ed.), Vol. I, Part 1, pp. 1–33. Masson, Paris.

Dawbin, W. H. (1962). The tuatara, New Zealands's ancient reptile. *Aust. Nat. Hist.* **14**, 32–34.

Delamare-Deboutteville, C., and Botosanéanu, L. (1970). "Formes primitives vivantes." Hermann, Paris.

Ford, E. B. (1971). "Ecological Genetics," 3rd Ed. Chapman and Hall, London.

Garstang, W. (1929). The morphology of the Tunicates and its bearings on the phylogeny of the Chordata. *Q. J. Microsc. Sci.* **72**, 51–187.

Gaskell, W. H. (1908). "The Origin of Vertebrates." Longmans, Green, London.

Grassé, P.-P. (1969). La classification de la faune terrestre. *In* "La Vie des Animaux," Vol. I, "Le peuplement de la terre," pp. 64–82. Larousse, Paris.

Haldane, J. B. S. (1956). The theory of selection for melanism in Lepidoptera. *Proc. R. Soc.*, *Ser. B* **145**, 303–306.

Jarvik, E. (1960). "Théories de l'évolution des Vertébrés." Masson, Paris.

Jeannel, R. (1944). "Les fossiles vivants des cavernes." Gallimard, Paris.

Jeannel, R. (1949a). Classification et phylogénie des Insectes. *In* "Traité de Zoologie" (P.-P. Grassé, ed.), Vol. IX, 3–85. Masson, Paris.

Jeannel, R. (1949b). Évolution et géonémie des Insectes. *In* "Traité de Zoologie" (P.-P. Grassé, ed.), Vol. IX, pp. 86–110. Masson, Paris.

Jordan, A. (1854). "Diagnoses d'Espèces Nouvelles ou Méconnues pour Servir à une Flaroe Réformée de la France et des Comtreés Voisines."

Jordan, A. (1873). Remarques sur le Fait de l'existence en société, a l'État Sauvage, des Espèces Végétales Affines et sue d'Autres Faits Relatifs à la Question de l'Espèce. *Assoc. Avanc. Sci.* Lyon, 2d session, p. 488.

Kelner-Pillault, S. (1969). Abeilles fossiles, ancêtres des Apides sociaux. *Proc. Union Int. Etude Insectes Soc. Bern, 6th, Congr.* pp. 85–93.

Kozhov, M. (1968). "Lake Baïkal and its Life." Junk, The Hague.

Martynov, A. B. (1938). Étude sur l'histoire géologique et la phylogénie des ordres des

Insectes Ptérygotes. I.—Palaeoptera, Neoptera, Polyneoptera. 150 p. *Tr. Paléontol. Inst. Akad. Nauk SSSR* **7**, Part 4.

Mastermann, A. T. (1897). On the notochord of *Cephalodiscus*. *Zool. Anz.* **20**, 443–450.

Mayr, E. (1945). "Systematics and the Origin of Species." Columbia Univ. Press, New York.

Millot, J., and Anthony, J. (1972). La glande post-anale de *Latimeria*. *Ann. Sci. Nat. Zool. Biol. Anim.* **14**, 305–318.

Millot, J., and Anthony, J. (1973). L'appareil excréteur de *Latimeria chalumnae* Smith (Poisson cœlacanthide). *Ann. Sci. Nat., Zool. Biol. Anim.* **15**, 293–328.

Moret, L. (1952). Embranchement des Spongiaires. *In* "Traité de Paléontologie" (J. Piveteau, ed.), Vol. I, pp. 333–374. Masson, Paris.

Murray, J., and Clarke, B. (1966). The inheritance of polymorphic shell characters in *Partula* (Gastropoda). *Genetics* **54**, 1261–1277.

Patten, W. (1890). On the origin of vertebrates from arachnids. *Q. J. Microsc. Sci.* **31**, 317–378.

Portmann, A. (1972). La cérébralisation des Mammifères. *In* "Traité de Zoologie" (P.-P. Grassé, ed.), Vol. XVI, Part 4, pp. 385–417. Masson, Paris.

Romer, A. S. (1963). "The Vertebrate Body" (3d ed.). Saunders Co, Philadelphia, Pennsylvania.

Stanier, R. Y., Doudoroff, M., Delberg, E. A. (1966). "Précis de Microbiologie." Masson, Paris.

Vallois, H. V. (1972). Le gisement et le squelette de Saint-Germain la Rivière. 3ᵉ partie. Anthropologie. *Arch. Inst. Paleontol. Hum. (Paris), Mem.* **34**, 45–112.

Vandel, A. (1972). La répartition des Oniscoïdes (Crustacés, Isopodes terrestres) et la dérive des continents. *C.R. Acad. Sci., Sér. D.* **275**, 2069–2072.

Welch, D'A. A. (1938). Distribution and variation of *Achatinella mustelina* in the Waianae Mountains, Oahu. *Berenice P. Bishop Mus. Bull.* **152**, 1–164, 671–684.

Chapter IV Evolution and Chance

Bogdanski, C. (1972). Contribution à l'étude des lois élémentaires de l'évolution des étres vivants. *C.R. Acad. Sci., Ser. D.* **274**, 1546–1549.

Bogdanski, C. (1972). Les êtres vivants comme une des principales classes des systèmes autorégulateurs naturels. *Bull. Biol. Fr. Belg.* **106**, 4–26.

Darwin, C. (1859). "Origin of the Species." John Murray, London.

Darwin, F. (1888). "Vie et Correspondance de Charles Darwin." (H. C. de Variguy, trans. from English), 2ᵈ vol., Reinwald, Paris.

Gabe, M et H. Saint-Girons (1967). Données histologique sur le tégument et les glandes épidermoides des Lepidosauriens. *Acta Anat.* **67**, 571–594.

Lamarck, J. B. Monnet, Chevalier de (1809). "Philosophie zoologique." (1873, 2d Ed.) Paris.

Ivanoff, (A.) (1953). "Les aberrations de l'œil." Revue d'Optique, édit. Paris.

Mukai, T. (1964). The genetic structure of natural populations of *Drosophila melanogaster*. I. Spontaneous mutation rate of polygenes controlling viability. *Genetics* **50**, 1–19.

Oparin, A. I. (1964). "The Chemical Origin of Life," (Ann Synge, trans.), Charles Thomas, Springfield, Illinois.

Oparin, A. I. (1967). État actuel du problème de l'origine de la vie et ses perspectives. *In* "Biogenèse" (J.-A. Thomas, ed.), pp. 17–29. Masson, Paris.

Poincaré, H. (1912b). "Science et Méthode." Flammarion, Paris.

Salet, G. (1972). "Hasard et certitude." Scient. Saint-Edme, Paris.

Strickberger, M. W. (1968). "Genetics." Macmillan, New York.

Waddington, C. H. (1965). "The Strategy of Genes," Allen & Unwin, London.
Wintrebert, P. (1962). "Le Vivant Créateur de son Évolution." Masson, Paris.

Chapter V Evolution and Natural Selection

Cain, A. J., and Currey, J. D. (1963). Area effects in *Cepaea*. *Philos. Trans. R. Soc., London, Ser. B* **246**, 1–81.
Cain, A. J., and Currey, J. D. (1968a). Climate and selection of banding morphs in *Cepaea* from the climate optimum to the present day. *Philos. Trans. R. Soc. London, Ser. B* **253**, 483–498.
Cain, A. J., and Currey, J. D. (1968b). Ecogenetics of a population of *Cepaea nemoralis* subjects to strong effects. *Philos. Trans. R. Soc. London, Ser. B* **253**, 447–482.
Cain, A. J., and Sheppard, P. M. (1950). Selection in the polymorphic land snail *Cepaea nemoralis*. *Heredity* **4**, 275–294.
Cain, A. J., and Sheppard, P. M. (1952). The effects of natural selection on body colour in the land snail *Cepaea nemoralis*. *Heredity* **6**, 217–231.
Cain, A. J., and Sheppard, P. M. (1954). Natural selection in *Cepaea*. *Genetics* **39**, 89–116.
Callot, G. (1958). Cited by Roman and Pichot (1972).
Cochran, D. G., Grayson, J. M., and Levitan, M. (1952). Chromosomal and cytoplasmic factors in transmission of DDT resistance in the German cockroach. *J. Econ. Entomol.* **45**, 997–1001.
Cuénat, L. and Tétry, A. (1951). "L'Evalution Biologique." Masson, Paris.
Darwin, C., and Wallace, A. R. (1958). "Evolution by Natural Selection." (1958 Ed. with a foreword by Sir Gavin de Beer. Cambridge Univ. Press, London and New York.)
Diver, C. (1929). Fossil records of Mendelian mutants. *Nature (London)* **124**, 183.
Dobzhansky, T. (1962). "Mankind Evolution." Yale University Press, New Haven, Conn.
Fisher, R. A. (1920). The measurement of selective intensity. *Proc. R. Soc., Ser. B* **121**, 58–62.
Fisher, R. A. (1930). "The Genetical Theory of Natural Selection." Clarendon Press, Oxford.
Ford, E. B. (1971). "Ecological Genetics," 410 pp. Chapman & Hall, London.
Haldane, J. B. S. (1924). A mathematical theory of natural and artificial selection. Part V to VIII. *Proc. Cambridge Philos. Soc.* **1**, 158–163, 363–373, 607–615, 838–844; **26**, 220–230; **27**, 131–136, 137–142.
Haldane, J. B. S. (1932). "The Causes of Evolution." Longmans and Green, New York.
Haldane, J. B. S. (1955). The measurement of natural selection. *Atti Congr. Int. Genet. Cariologia, 9th, Florence* Suppl., pp. 480–487.
Haldane, J. B. S. (1956). The theory of selection for melanism in Lepidoptera. *Proc. R. Soc., Ser. B* **145**, 303–306.
Kimura, M., and Ohta, T. (1971). "Theoretical Aspects of Population Genetics." Princeton Univ. Press, Princeton, New Jersey.
Lamotte, M. (1951). Recherches sur la structure génétique des populations naturelles de *Cepaea nemoralis* L. *Bull. Biol. Fr. Belg., Suppl.* **35**, 1–239.
Lamotte, M. (1966). Les facteurs de la diversité du polymorphisme dans les populations naturelles de *Cepaea nemoralis*. *Lav. Soc. Malacol. Ital.* **3**, 33–73.
Lederberg, J. and E. L. Tatum (1946). Gene Recombination in *Escherichia coli*. *Nature* **158**, 558.
Malthus, T. R. (1798). "Essay on the Principle of Population." "Essai sur le principe de population" (Fr. transl. by M. M. P. Prévost and G. Prévost), 2nd Ed. Guillaumin, Paris, 1852.

Moore, J. A. (1946). Incipient intraspecific isolating mechanism in *Rana pipiens*. *Genetics* **31**, 304–326.

Moore, J. A. (1949). Geographic variation of adaptative characters in *Rana pipiens* Schreber. *Evolution* **3**, 1–24.

Perrier, E. (1896). "La Philosophie Zoologique avant Darwin." F. Y. Alcan, Paris.

Ramade, F. (1967). Contribution à l'étude du mode d'action de certains insecticides de synthèse plus particulièrement du lindane et des phénomènes de résistance à ces composés chez *Musca domestica*. *Ann. Inst. Nat. Agron.* (Paris) **5**, 1–268.

Roman, E., and Pichot, J. (1972). Variations, pendant un quart de siècle, de la sensibilité au DDT des larves de *Culex pipiens* autogène à Lyon et dans ses environs. *C. R. Acad. Sci., Sér. D* **274**, 1187–1189.

Rothschild, Lord W. (1907). "Extinct Birds." London pp. 1–244 (name of the editor unknown). p. 344.

Strickland, H. E., and Melville, A. G. (1848). "The Dodo and it's Kindred." Reeve, Benhan and Reeve, London.

Teissier, G. (1943). Apparition et fixation d'un gène mutant dans une population stationnaire de Drosophiles. *C.R. Acad. Sci.* **216**, 88.

Teissier, G. (1958). "Titres et travaux scientifiques." Prieur and Robin, Paris.

Teissier, G. (1962). Supplément aux Titres et Travaux de G. T. Robin et Mareuge, Paris.

Wallace, A. R. (1908). *Note on the passage of Malthus' "Principle of Population," which suggested the idea of natural selection to Darwin and myself.* "Darwin and Wallace celebration held on Thursday 1 July, 1908," pp. 111–118. London.

Wright, S. (1969). "Evolution and the Genetics of Populations," 3 vols. Univ. of Chicago Press, Chicago, Illinois.

Chapter VI Evolution and Adaptation

Bernardin de Saint-Pierre, J.-H. (1784). Études de la Nature. *In* "Oeuvres complètes," Vol. II. André, Paris, 1823.

Bonnier, G. (1890). Cultures expérimentales dans les Alpes et les Pyrénées. *Rev. Gen. Bot.* **2**, 513.

Bonnier, G. (1894a). Recherches expérimentales sur l'adaptation des plantes au climat alpin. *Ann. Sci. Nat., Bot. Biol. Vég.* **20**, 217–360.

Bonnier, G. (1894b). Les plantes arctiques comparées aux mêmes espèces des Alpes et des Pyrénées. *Rev. Gen. Bot.* **6**, 505–527.

Burgin-Wyss U. (1961). Die Ruckenhänge von *Trinchesia coerulea* (Montagu). Eine morphologische Studie über Farbmuster bei Nudibranchiern. *Rev. Suisse Zool.* **68**, 461–582.

Cain, A. J., and Currey, J. D. (1968). Climate and selection of banding morphs in *Cepaea* from the climate optimum to the present day. *Philos. Trans. R. Soc., London, Ser. B* **253**, 484–498.

Cain, A. J. and Sheppard, P. M. (1950). Selection in the polymorphic land snail *Cepaea nemoralis*. *Heredity*, **4**, 275–294.

Dechazeaux, C. (1960) Bivalves Fossiles. *In* "Traité de Zoologie," Vol. V, Book 2, pp. 2134–2164. Masson, Paris.

Deleurance-Glaçon, S. (1963). Recherches sur les Coléoptères troglobies de la sous-famille des Bathysciinae. *Ann. Sci. Nat., Zool. Biol. Anim.* **5**, 1–172.

Dobzhansky, T., and Boesiger, E. (1968). "Essais sur l'évolution." Masson, Paris.

Ford, E. B. (1971). "Ecological Genetics." Chapman and Hall, London.

Gertrude, M. T. (1937). Métabolisme et morphogenèse en milieu aquatique. *Rev. Gen. Bot.* **49**, 161.

Giard, A. (1904). Controverses Transformistes. Naud, Paris.

Glaçon, S. (1953). Sur le cycle évolutif d'un Coléoptère troglobie, *Speonomus longicornis S. C.R. Acad. Sci. Sér D*, **232**, 1027–1029.

Grant, V. (1962). "The Origin of Adaptations," 606 pp. Columbia Univ. Press, New York.

Grassé, P.-P. and Noirot, C. (1951). La sociotomie, migration, et fragmentation de la termitière chez *Anoplotermes* et *Trinervitermes*. *Behavior*, **3**, 153–160.

Huxley, J. (1945). "Evolution: The Modern Synthesis." Allen & Unwin. First ed. London.

Jeannel, R. (1952). Psélaphides recueillis par N. Lelup au Congo belge. IV Faune de l'Itombwee et du Rugege. *Ann. Mus. R. Congo Belge, Sci. Zool.* **11**, 295 pp.

Jeannel, R. (1953). Origine et répartition des Psélaphides de l'Afrique intertropicale. *Trans. Int. Congr. Entomol. 9th, Amsterdam.* Vol. II, 154–156 (1951).

Jeannel, R., and Leleup, N. (1952). L'évolution souterraine dans la région méditerranéenne et sur les montagnes du Kivu. *Notes Biospeol.* **7**, 7–13.

Kepner, W. A., Gregory, W. C., and Porter, R. J. (1935). The manipulation of the nematocysts de *Chlorohydra* by *Microstomum*. *Zool. Anz.* **121**, 204–220.

Kepner, W. A., and Nuttycombe, J. W. (1929). Further studies upon the nematocysts of *Microstomum caudatum. Biol. Bull.* **57**, 69–78.

Lamotte, M. (1951). Recherches sur la structure génétique des populations naturelles de *Cepaea nemoralis. L. Bull. Biol. Fr. Belg., Suppl.* **35**, 1–239.

Lamotte, M. (1959). Polymorphism of natural populations of *Cepaea nemoralis. Cold Spring Harbor Symp. Quant. Biol.* **24**, 65–86.

Lamotte, M. (1966). Les facteurs de la diversité du polymorphisme dans les populations naturelles de *Cepaea nemoralis. Lav. Soc. Malacol. Ital.* **3**, 33–73.

Lederberg, J., and Lederberg, E. M. (1952). Replica plating and indirect selection of bacterial mutants. *J. Bacteriol.* **63**, 399–406.

Luria, S. E. and Delbrück, M. (1943). Mutations of bacteria from virus sensivity to virus resistance. *Genetics* **28**, 491–511.

Matisse, G. (1942). "L'éternelle Illusion." Editions d'Art et d'Historie, Paris.

Matsakis, J. (1962). Contribution à l'étude du développement prostembryonnaire et de l'évolution de la forme chez quelques Crustacés Isopodes. *Bull. Biol. Fr. Belg.* **96**, 531–691.

Meixner, J. (1923). Ueber die Kleptokniden von *Microstomum lineare* (Müll.). *Biol. Zentralbl.* **43**, 559–573.

Molliard, M. (1907). Action Morphogénétique de quelques Substances Organiques sur les Végétaux supérieurs. *Rev. Gén. Bot.* **19**, 241–290.

Newman, B. H. (1968). Raking up the dead. *Nature (London)* **220**, 123.

Piéron, H. (1945). "La Sensation Guide de Vie." Gallimard, Paris.

Plate, L. (1913). "Selektionsprinzip and Probleme der Artbildung," 4th Ed. Leipzig and Berlin.

Rabaud, E. (1922). "L'adaptation et l'évolution," 284 pp. Chiron, Paris.

Rabaud, E. (1932–1934). "Zoologie biologique." 2 vol. Gauthier-Villars, Paris.

Sencer, D. J. (1971). *Annu. Rev. Microbiol.* 465–486.

Thinès, G. (1969). "L'évolution régressive des Poissons cavernicoles et abyssaux." Masson, Paris.

Vandel, A. (1964). "La biospéologie. La biologie des animaux cavernicoles," Gauthier-Villars, Paris.

Weismann, A. (1885). La continuité du plasma germinatif comme base d'une théorie de l'hérédité. *In* "Essais sur l'hérédité et la sélection naturelle" (transl. from Ger. by H. de Varigny). Reinwald, Paris, 1892.

Weismann, A. (1886). La régression dans la nature. Conférence donnée à l'Akademische

Gesellschaft de Fribourg-en-Brisgau. *In* "Essais sur l'hérédité et la sélection naturelle" (transl. from Ger. by H. de Varigny), pp. 381–409. Reinwald, Paris, 1892.

Weismann, A. (1888). La prétendue transmission héréditaire des mutations. Conférence donnée à la Réunion des Naturalistes allemands, Cologne. *In* "Essais sur l'hérédité et la sélection naturelle" (transl. from Ger. by H. de Varigny). Reinwald, Paris, 1892.

White, M. J. D. (1973). "Animal Cytology and Evolution." Cambridge Univ. Press, 3[d] edit. London and New York.

Chapter VII Evolution and Necessity

Boutroux, E. (1874). "De la contingence des lois de la nature." F. Alcan, Paris.

Laporte, J. (1941). "L'idée de nécessité," 157 pp. F. Alcan, Paris.

Monod, J. (1970). "Le hasard et la nécesité." Le Seuil, Paris.

Le Molliard, M. (1907). Action morphogénétique de quelques substances organiques sur les Végétaux supérieurs. *Rev. Gén. Bot.* **19**, 241–290.

Huxley, J. (1942). Evolutionary progress. *In* "Evolution, the Modern Synthesis," pp. 556–578. Allen & Unwin, London.

Rabaud, E. (1953). "Le hasard et la vie des espèces." Flammarion, Paris.

Chapter VIII Activities of the Genes in Relation to Evolution.

Avery, D. T., Mac Leod, C. M., and Mc Carty, M. (1944). Studies on the chemical nature of the substance inducing transformation of pneumococcal typhus. Induction of transformation by a desoxyribonucleic fraction isolated from *Pneumococcus* type III. *J. Exp. Med.* **79**, 137–158.

Beadle, G. W., and Tatum, E. L. (1941). Genetic control of biochemical reactions in Neurospora. *Proc. Natl. Acad. Sci. U.S.A.* **27**, 499–506.

Beermann, W. (1972). "Developmental Studies on Giant Chromosomes. Results and Problems in Cell Differentiation," 227 pp. Springer-Verlag, Berlin and New York.

Bolton, E. T. and McCarthy, B. J. (1962). A general method of isolation of RNA complementary to DNA. *Proc. Natl. Acad. Sci. U.S.A.* **48**, 1390–1395.

Borchardt, E. (1927). Beitrag zur heteromorphen Regeneration bei *Dixippus morosus*. *Arch. Entwicklungsmech. Org.* **110**, 366–394.

Britten, R. J. and Davidson, E. H. (1969). Gene regulation for higher cells: a theory. *Science* **165**, 347–357.

Britten, R. J. and Kohne, D. E. (1970). Repeated segments of DNA. *Sci. Am.* **222**, 24–31.

Commoner, B. (1968). Failure of the Watson-Crick theory as chemical explanation of inheritance. *Nature, (London)*, **227**, 561–563.

Cuénot, L. (1921a). Régénération de pattes à la place d'antennes sectionnées chez un Phasme. *C.R. Acad. Sci.* **172**, 949–951.

Cuénot, L. (1921b). Sur les différents modes de régénération des antennes chez le Phasme *Carausius morosus*. *C.R. Acad. Sci.* **172**, 1009–1011.

Davidson, J. N. (1972). "The Biochemistry of the Nucleic Acids," 7th Ed. Chapman & Hall, London.

Dawydoff, C. (1942). La régénération créatrice des Némertes. *Bull. Biol. Fr. Belg.* **76**, 58–141.

de Grouchy, J. *et al.* (1972). Evolutions caryotypiques de l'Homme et du Chimpanzé. *Ann. Genet.* **15**, 79.

Dobzhansky, T., and Pavlovsky, O. (1962). A comparative study of the chromosomes in the incipient species of the *Drosophila paulistorum* complex. *Chromosoma* **13**, 196–218.

Dollo, L. (1893). Les lois de l'évolution. *Bull. Soc. Belg. Geol.* **7**, 164–166.

Driesch, H. (1921). "La philosophie de l'organisme" (M. Kollman, transl.). Rivière, Paris.

Feulgen R. and Rosenbeck (1924). Mikroskopisch-chemischer Nachweis einer Núclein-säure vom Typus der Thymolnucleinsäure und die darauf beruhende elektive Far-bung von Zellkern in mikroskopischen Präparaten. *Hoppe Seyler's Z. Physiol. Chem.*, **135**.

Gallo, R. C. (1971). Reverse transcriptase, the DNA transcriptase of oncogenic RNA viruses. *Nature (London)* **234**, 194–198.

Giard, A. (1897). Sur les régénérations hypotypiques. *C.R. Soc. Biol.* **4**, 315–317.

Goodman, M. (1961). The role of immunochemical differences in the phyletic development of human behavior. *Hum. Biol.* **33**, 131–162.

Hatanaka, M., Kakefuda, T., *et al.* (1971). Cytoplasmic DNA synthesis induced by RNA tumor viruses. *Proc. Natl. Acad. Sci. U.S.A.* **68**, 1844–1847.

Herbst, C. (1896). Ueber die Regeneration von antennenähnlichen Organen an Stelle von Augen. *Arch. Entwicklungsmech. Org.* **2**, 544–563.

Jacob, F., and Monod, J. (1961). Genetic regulatory mechanism in the synthesis of pro-teins. *J. Mol. Biol.* **3**, 318–356.

Karpechenko, 0. (1928). Polyploid hybrids of *Raphanus sativus* x *Brassica oleracea* L. *Bull. appl. Bot.* Leningrad, **77**, 305–340 (English summary).

Kohne, D. E. (1970). Evolution of higher organism DNA. *Q. Rev. Biophys.* **3**, 327.

Lederberg, J., and Tatum, E. L. (1946). Gene recombination in *Escherichia coli*. *Nature (London)* **158**, 558.

Lejeune, G., Dutrellaux, B., and Grouchy, J., de (1970). Reciprocal Translocation in Human Population. A Preliminary Analysis. *In* "Population Genetics," Edinburgh Univer-sity Press, Edinburgh, Scotland.

Loeb, J. (1891). "Untersuchungen zur physiologischen Morphologie der Thiere," Vol. I, "Heteromorphosis." Wurzburg.

Lorković, Z. (1949). Chromosomen-Vervielfachung bei Schmetterlingen und ein neuer Fall fünffacher Zahl. *Rev. Suisse Zool.* **56**, 243–249.

Lubischev, A. (1925). On the nature of hereditary factors (in Russ. with Engl. summary). *Izv. Permsk. Biol. Nauchno-Issled. Inst.* **4**.

MacGregor, H. C. and Uzzell, T. M., Jr. (1964). Gynogenesis in Salamanders related to *Ambystoma jeffersonianum*. *Science* (N.Y.) **143** (1043–1045).

Matthey, R. (1969). Les Chromosomes et l'évolution chromosomique des Mammifères. *In* "Traité de Zoologie" (P.-P. Grassé, ed.), Vol. XVI, Part VI, pp. 855–909. Masson, Paris.

Matthey, R. (1970). Les chromosomes des Reptiles. *In* "Traité de Zoologie" (P.-P. Grassé, ed.), Vol. XIV, Part 3, pp. 829–858. Masson, Paris.

Matthey, R., and Jotterand, M. (1972). Deux espèces cryptiques sont confondues sous le nom de *Gerbillus gracilis* Th. (Gerbillinae). *Rev. Suisse Zool.* **79**, 1104–1105.

Monod, J., and Jacob, F. (1961). Teleonomic mechanism in cellular metabolism, growth and differenciation. *Cold Spring Harbor Symp. Quant. Biol.* **26**, 389–401.

Nirenberg, M. W., Jones, O. W., Leder, P., Clark, B. F. C., Sly, W. S., and Pestka, S. (1963). On the coding of genetic information. *Cold Spring Harbor Symp. Quant. Biol.* **28**, 549–563.

Ohno, S. (1970). "Evolution by Gene Duplication." Springer-Verlag, Berlin and New York.

Schildkraut, C., Marmur, J., and Doty, P. (1961). The formation hybrid DNA molecules and their use in studies of DNA homologies. *J. Mol. Biol.* **5**, 595–617.

Speyer, J. E., Lengyei, P., Basilio, A., Wahba, A. J. Gardner, R. S., and Ochoa, S. (1963). Synthesis and structure of macromolécules. *Cold Spring Harbor Symp. Quant. Biol.* **28**, 559–570.

Sutton, W. S. (1903). The chromosomes in heredity. *Biol. Bull.* **4**, 231–251.

Temin, H. M. (1971). Mechanisms of cell transformation by RNA tumor viruses. *Annu. Rev. Microbiol.* **25**, 609–648.

Temin, H. M., and Mizutani, S. (1970). RNA-dependent DNA polymerase in virions of Rous sarcoma virus. *Nature (London)* **266**, 1211–1213.

Turleau, C., and de Grouchy, J. (1972). Caryotypes de l'Homme et du Chimpanzé. Comparaison de la topographie des bandes. Mécanismes évolutifs possibles. *C.R. Acad. Sci. Sér. D*, **274**, 2355–2358.

Turleau, C., *et al.* (1972). Phylogénie chromosomique de l'Homme et des Primates hominiens *(Pan troglodytes, Gorilla gorilla, Pongo pygmaeus). Ann. Genet.* **15**, 225–237.

Waring, M., and Britten, R. J. (1966). Nucleotide sequence repetition: a rapidly reassociating fraction of mouse DNA. *Science* **154**, 791–794.

Wilson, E. B. (1925). "The Cell in Development and Heredity," Ch. XII. Macmillan, New York.

Chapter IX A New Interpretation of Evolutionary Phenomena.

Aaronson, S. A., Parks W. P., Scolnick, E. M., and Todaro G. J. (1971). Antibody to the RNA-dependent, DNA polymerase of mammalian C-type RNA tumor viruses. *Proc. Nat. Acad. Sci. U.S.A.* **68**, 920–924.

Anderson, N. G. (1970). Evolutionary significance of virus infection. *Nature (London)* **227**, 1346–1347.

Baltimore, D. (1970). Viral RNA dependent DNA polymerase. *Nature (London)* **226**, 1209–1211.

Barrell, B. G., Air, G. M., and Hutchinson, C. A., III (1976). Overlapping genes in bacteriophage φX174. *Nature (London)* **264**, 34–41.

Bataillon, E., and Tchou-Su (1929). Analyse de la fécondation chez les Batraciens par l'hybridation et la polyspermie physiologique. *Arch. Entwicklungsmech. Org.* **115**, 780–824.

Beadle, G. W., and Tatum, E. C. (1941). Genetic control of biochemical reactions in *Neurospora. Proc. Natl. Acad. Sci. U.S.A.* **27**, 449–506.

Beermann, W. (1972). "Developmental Studies on Giant Chromosomes. Results and Problems in Cell Differentiation." Springer, Berlin.

Beermann, W., and Clever, V. (1964). Chromosome puffs. *Sci. Am.* **180**, 50–58.

Beljanski, M. (1972). Synthèse *in vitro* de l'ADN sur une matrice d'ARN par une transcriptase d'*Escherichia coli. C. R. Acad. Sci., Ser. D.* **274**, 2801–2804.

Beljanski, M., Beljanski, M. S., Manigault, P. and Bourgarel, P. (1972). Transformation of *Agrobacterium tumefaciens* into a non-oncogenic species by an *Escherichia coli* RNA. *Proc. Natl. Acad. Sci. U.S.A.* **69**, 191–195.

Beljanski, M., Beljanski, M. S., Plawecki, M., and Bourgarel, P. (1975). *C.R. Acad. Sci., Sér. D*, **280**, 363–366.

Benzer, S. (1959). On the topology of genetic fine structure. *Proc. Natl. Acad. Sci. U.S.A.* **45**, 1607–1612.

Benzer, S. (1961). On the topography of genetic fine structure. *Proc. Natl. Acad. Sci. U.S.A.* **47**, 403–418.

Boveri, T. (1902). "Das Problem der Befruchtung." G. Fischer, Jena.

Boveri, T. (1914). Ueber die Charactere von Echiniden-Bastardlarven bei verschiedenem Mengenverhältnis mütterlicher und väterlicher Substanzen. *Verhandl. Phys. Med. Gesellsch. Würzburg* **43**, 502–520.

Boveri, T., and Hogue, M. (1909). Ueber die Möglichkeit Ascariseir zur Teilung in zwei gleichvertige Blastomeren zu veranlassen. *Verhandl. Phys. Med. Gesellsch. Würzburg* **15**, 403–420.

Brachet, A. (1922). Recherches sur la fécondation prématurée de l'œuf d'Oursin *(Paracentrotus lividus)*. *Arch. Biol.* **32**, 205–248.

Braun, A. C., and Wood, H. N. (1966). On the inhibition of tumor inception in the Crown-gall disease with the use of ribonuclease A. *Proc. Natl. Acad. Sci. U.S.A.* **56**, 1417–1422.

Cavalieri, L. F. (1963). Nucleic acids and information transfer. *J. Cell. Comp. Physiol.* **62**, Suppl. 1, 111–122.

Chambers, R. (1917a). Microdissection studies. *Am. J. Physiol.* **161**, 520–540.

Chambers, R. (1917b). Microdissection studies. *J. Exp. Zool.* **25**, 412–428.

Chambers, R. (1921). Microdissection studies. *Biol. Bull.* **42**, 1021–1030.

Champy, C. (1924). "Sexualité et hormones," 376 pp. Doin, Paris.

Chargaff, E. (1950). Chemical specificity of nucleic acids and mecanism of their enzymatic degradation. *Experientia* **6**, 201–209.

Chargaff, E. (1965). "Developmental and Metabolic Control Mechanisms and Neoplasia." Williams & Wilkins, Baltimore, Maryland.

Commoner, B. (1968). Failure of the Watson-Crick theory as chemical explanation of inheritance. *Nature (London)* **220**, 334–340.

Crick, F. (1970). Central dogma of molecular biology. *Nature (London)* **227**, 561–563.

Darwin, C. (1868). "Origin of Species". John Murray, London.

Darwin, Ch. (1868). "Variation of Animals and Plants under Domestication." London.

Davidson, J. N. (1972). "The Biochemistry of the Nucleic Acids," 7th ed., Chapman and Hall, London.

Dawydoff, C. (1942). La régénération créatrice chez les Némertes. *Bull. Biol. Fr. Belg.* **76**, 58–140.

Driesch, H. (1921). "La philosophie de l'organisme" (M. Kollmann, transl.). Rivière, Paris. ("Philosophie des Organismen," 1909.)

Ford, E. B. (1971). "Ecological Genetics." 3rd ed., Chapman et Hall, London.

Gabe, M., and Saint-Girouns, H. (1965). Contribution à la morphologie comparée du cloaque et des glandes épidermoïdes de la région cloacale, chez les Lépidosauriens. *Mem. Mus. Nat. Hist. Nat. Sér. A Zool.* **33**, 151–292.

Gabe, M., and Saint-Girons, H. (1967). Données histologiques sur le tégument et les glandes épidermoïdes des Lépidosauriens. *Acta Anat.* **67**, 571–594.

Gegenbaur, C. (1886). "Zur Kenntniss der Mammarorgane des Monotremen." Engelmann, Leipzig.

Gerassimova. Personal communication.

Godlewski, E. (1906). Untersuchungen über die Bastardierung der Echiniden- und Crinoidenfamilie. *Arch. Entwicklungsmech. Org.* **20**, 312–340.

Godlewski, E. (1911). Kombination der heterogenen Befruchtung mit der künstlichen Parthenogenesis. *Arch. Entwicklungsmech. Org.* **33**, 180–213.

Goulian, M., Kornberg, A., and Sinsheimer, R. L. (1967). Enzymatic synthesis of DNA. XXIV. Synthesis of infections phage X 174 DNA. *Proc. Natl. Acad. Sci. U.S.A.* **58**, 2321–2328.

Grassé, P.-P. (1957). Ultrastructure, polarité et reproduction de l'appareil de Golgi. *C.R. Acad. Sci. Sér D*, **245**, 1278–1281.

Grassé, P.-P. (1961). La reproduction par induction du blépharoplaste et du flagelle de *Trypanosoma equiperdum* (Flagellé protomonadine). *C.R. Acad. Sci. Sér D,* **252,** 3917–3921.

Grassé, P.-P., and Faure, A. (1935). La reproduction de l'appareil parabasal de *Trichomonas caviae* (Dav). *C.R. Acad. Sci.* **200,** 1493–1495.

Gurdon, J. B. (1963). Nuclear transplantation in Amphibia and the importance of stable nuclear changes in promoting cellular differentiation. *Q. Rev. Biol.* **38,** 54–78.

Haeckel, E. (1882). "Les preuves du transformisme, réponse à Virchow" (J. Soury, transl.), 2nd Ed. Baillière, Paris. (Orig. Germ. Ed.: 1877.)

Haldane, J. B. S. (1959). The theory of natural selection to day. *Nature, London* **183,** 710–713.

Hatanaka, M., Kakefuda, T., *et al.* (1971). Cytoplasmic DNA synthesis induced by RNA tumor viruses. *Proc. Natl. Acad. Sci. U.S.A.* **68,** 1844–1847.

Hill, M., and Hillova, J. (1972). Virus recovery in chicken cells tested with Rous-sarcoma cell DNA. *Nature (London), New Biol.* **237,** 35.

Johansen, W. (1911). The genotype conception of heredity. *Am. Nat.* **45,** 129–159.

Jordan, A. (1846–1848). Observations sur plusieurs plantes nouvelles, rares ou critiques de la France. *Mem. Soc. Sci. Nat. Lyon.*

Jordan, A. (1864). Diagnoses d'espèces nouvelles ou méconnues pour servir à une flore réformée de la France et des contrées voisines.

Kacian, D. L., Spiegelman, S., *et al.* (1972). *In vitro* synthesis of DNA components of human genes for globins. *Nature (London), New Biol.* **235,** 167–169.

Kohne, D. E. (1970). Evolution of higher organism DNA. *Q. Rev. Biophys.* **3,** 327–375.

Kornberg, A. (1960). Biologic synthesis of desoxyribonucleic acid. *Science* **131,** 1503–1608.

Kornberg, A. (1962). "Enzymatic Synthesis of DNA." Wiley, New York.

Kornberg, A. (1969). Active center of DNA polymerase. *Science* **163,** 1410–1418.

Lameere, A. (1915). Les caractères sexuels secondaires des Prionides. *Bull. Biol. Fr. Belg.* **49,** 1–14.

Mach, B., Reich, E., and Tatum, E. L. (1963). Separation of the biosynthesis of the antibiotic polypeptide tyrosidine from protein biosynthesis. *Proc. Natl. Acad. Sci. U.S.A.* **50,** 175.

Merril, C. R., Grier, M. R., Petricciani, J. C. (1971). Bacterial versus gene expression in human. *Nature* (G.B.) **233,** 398–400.

Miller, O. L., Hamkalo, B. A., and Thomas, G. A. (1970). Visualization of bacterial genes in action. *Science* **169,** 392–395.

Mills, D. R., Kramer, F. R., and Spiegelman, S. (1973). Complete nucleotide sequence of a replicating RNA molecule. *Science* **180,** 916–927.

Monod, J. (1971). "Chance and Necessity." Knopf, New York.

Morgan, C. L. (1927). "Emergent Evolution." H. Holt, New York.

Morgan, T. H. (1911). An attempt to analyse the constitution of the chromosomes on the basis of sex-limited inheritance in *Drosophila. J. Exp. Zool.* **11,** 365–414.

Naegeli, von C. (1884). "Mechanisch-physiologische Theorie der Abstammungslehre," 822.

Ohno, S. (1970). "Evolution by Gene Duplication." Springer-Verlag, Berlin and New York.

Osborn, H. F. (1929). The Titanotheres of ancient Wyoming. *Monogr. U.S. Geol. Sur.,* No. 55, **1,** 1–701.

Penard, C. (1938). "Les infiniments petits dans leurs manifestations vitales." Georg, Genova.

Ross, J., Aviv, H., *et al.* (1972). *In vitro* synthesis of DNA complementary to purified rabbit globin mRNA. *Proc. Natl. Acad. Sci. U.S.A.* **69,** 264–268.

Roux, W. (1881). "Der Kampf der Teile im Organismus." *Arch. Entwicklungsmech. Org.* (1895) **1,** 137–437.

Sabol, S., Sillero, M. A. G., Iwasaki, K., and Ochoa, S. (1970). Purification and Properties of Initiation Factor F3. *Nature* **228**, 1269–1273.

Schmalhausen, I. (1949). "Factors of Evolution." Blakiston, Philadelphia, Pennsylvania.

Spemann, H. (1938). "Embryonic Development and Induction." Yale University Press.

Spiegelman, S., *et al.* (1970a). Characterization of the products of RNA directed DNA polymerases in oncogenic RNA viruses. *Nature (London)* **227**, 563–567.

Spiegelman, S., *et al.* (1970b). DNA directed DNA polymerase activity in oncogenic RNA viruses. *Nature (London)* **227**, 1029–1031.

Spiegelman, S., *et al.* (1970c). Synthetic DNA-RNA hybrids and RNA-RNA duplexes as templates for the polymerases of the oncogenic RNA viruses. *Nature (London)* **228**, 430–432.

Spurway, H. (1949). Remarks on Vavilov's law of homologous variation. *Ric. Sci., Suppl.,* Symposium **18**, 212–220, Rome.

Tavlitzki, J. (1972). Le code génétique. *Biologie génétique,* Monogr. Annu. Soc. Fr. Biol. Clin., 55–62.

Teissier, G., and Huxley, J. (1936). Terminologie et notation dans la description de la croissance relative. *C.R. Soc. Biol.* **121**, 934.

Temin, H. M. (1964). *Natl. Cancer Inst. Monogr.* **17**, 557–570.

Temin, H. M. (1971). Mechanism of cell transformation by RNA tumor virues. *Annu. Rev. Microbiol.* **25**, 609–648.

Temin, H. M., and Mizutani, S. (1970). RNA-dependent DNA polymerase in visions of Rous Sarcoma virus. *Nature (London)* **266**, 1211–1213.

Trilling, D. M., and Axelrod, D. (1970). Encapsidation of free host DNA by Simian virus-40: A Simian virus-40 pseudovirus. *Science* (U.S.A.) **168** (3928), 268–271.

Waddington, C. H. (1957). "Strategy of the Genes." Allen & Unwin, London.

Watson, J. D., and Crick, F. C. (1953). Molecular structure of nucleic acids. A structure of desoxyribose nucleic acids. *Nature (London)* **171**, 737–738.

Weismann, A. (1892). "Das Keimplasma. Eine Theorie der Vererbung." Jena.

Weismann, A. (1902): "On Germinal Selection as a Source of Definite Variation." Religion and Science Library, London. (Orig. Germ. Ed.: 1896.)

Weismann, A. (1902). "Vorträge über Deszendenztheorie," Vols. I and II. Jena.

Whyte, L. L. (1965). "Internal Factors in Evolution." Associated Book Publishers, London.

Wintrebert, P. (1962). "Le vivant créateur de son évolution." Masson, Paris.

Wittrock, 0. (1903). Cited by L. Cuénot (1921).

Wolff, E. (1971). Les grands problèmes posés par un petit Crapaud vivipare et accoucheur de Guinée. *Sciences (Paris)* Nos. 74–75, 32–38.

Woodger, J. H. (1959). "Studies on the Foundations of Genetics in the Axiomatic Method." North-Holland Publ., Amsterdam.

Author Index

Numbers in italics refer to the pages on which the complete references are listed.

A

Aaronson, S. A., 221, *274*
Air, G. M., 235, *274*
Alvarado, R., et. al., *262*
Anderson, N. G., 239, *274*
Anthony, J., *268*
Anthony, R., *262*
Avery, D. T., 183, *272*
Aviv, H., 222, *276*
Axelrod, 239

B

Bailey, D. W., 84, *267*
Baldwin, J. M., *262*
Baltimore, D., 221, *274*
Barghoorn, E. S., 11, 12, *264*
Barghusen, H. R., 42, 43, *265*
Barrell, B. G., 235, *274*
Basilio, A., 184, *273*
Bataillon, E., 209, *274*
Bateson, 255, 258
Bauchot, R., 64, *267*
Beadle, G. W., 186, 232, *272*, *274*
Beermann, W., 185, 218, *272*, *274*
Beljanski, M. S., 238, *274*
Beljansky, M., 222, *274*
Benzer, S., 233, *274*
Berg, E. S., *262*
Bergson, H., *262*

Bernardin de Saint-Pierre, J.-H., 133, *270*
Bhagavan et. al. (1966), 241
Boesiger, E., 171, *262*, *270*
Bogdanski, C., 93, *268*
Bolk, L., *262*
Bolton, E. T., 190, *272*
Bonnet, Charles, 216
Bonnier, G., 148, *270*
Borchardt, E., 197, *272*
Botosanéanu, L., 78, *267*
Boureau, E., *264*
Bourgarel, P., 222, 238, *274*
Boutroux, E., *272*
Boveri, T., 184, 220, *274*
Brachet, A., 220, *274*
Braun, A. C., *274*
Britten, R. J., 190, 191, *272*, *274*
Burgin-Wyss, 136, *270*

C

Cain, A. J., 114, 142, *269*, *270*
Callot, G., 120, *269*
Caullery, M., *262*
Cavalieri, L. F., 238, *275*
Chadefaud, M., *264*
Chaline, J., 71, *267*
Chambers, R., 220, *275*
Champy, C., 217, *275*
Chargaff, E., 231, *275*
Clark, B. F. C., 184, *273*
Clarke, B., 84, *268*

279

Subject Index

A

Acanthostega, 73
Achatinella, 63, 84
Achondroplastic dwarfism, 96
Acrididae, 116, 192
Acrochordus, 22
Acrocinus longimanus, 145
Actinistian crossopterygians, 73
Actinoids, 15
Actinomyces, 59
Actinomycin, 255
Actinopterygians, 63, 189
Aëdes aegypti, 120
Aeolians, 136
Aesculus carnea, 201
Aethia rossmoori, 68
Agnatha, 17, 60, 189
Agrobacterium tumefaciens, 222
Ainiktozoon, 17
Akaryotes, 11, 179
Albatross, 67
Allele, 100, 185, 255
Allelomorph, 255
Alligator, 189
Allophalomys pliocaenicus, 71
Alticamelids, 146
Amber, Baltic, 62
Amblyopsis, 157
Ambystoma laterale, 189; A. tigrinum, 189
Ameba, 178
Amino acids, 255
Ammonites, 63
Amphibians, 17, 32
  anuran, 189

lybyrinthodont stegocephalian, 74
  urodelan, 157, 189
Amphioxus, 17, 78
Amphipods, 86, 175
Amphisbenians, 21, 23
Amphiuma, 189
Anagale, 70, 83
Anaspida, 77
Anastosaurus, 141
Anchitherium, 50
Aneides lugubris, 136
Angström, 255
Anguimorphs, 21
Anguinidae, 21, 25
Anguis, 21, 23
Animal kingdom (classification), 248
Anisoscelis foliaceus, 145
Annelids, 4, 5, 17, 29; polychetes, 82
Anomia, 77
Anomodonts, 33, 45
Anthropomorphs, 64; pongids, 64
Antibodies, 134
Ant-lion, 161–162
Ants, 169
Anvil (bone), 52
Apathy, 15
Apis, 82
Apneumony, 136
Apocrine gland, 255
Apody, 22, 23, 255
Apterous diptera, 111
Apus gaillardi, 68; A. ignotus, 68
Arachnids, 157
Arca, 77
Archaeocete Cetacea, 145